SAFER SYSTEMS

Springer
*London*
*Berlin*
*Heidelberg*
*New York*
*Barcelona*
*Budapest*
*Hong Kong*
*Milan*
*Paris*
*Santa Clara*
*Singapore*
*Tokyo*

*Related titles:*

Directions in Safety-critical Systems
Proceedings of the First Safety-critical Systems Symposium, Bristol 1993
Redmill and Anderson (eds)
3-540-19817-2

Technology and Assessment of Safety-critical Systems
Proceedings of the Second Safety-critical Systems Symposium, Birmingham 1994
Redmill and Anderson (eds)
3-540-19859-8

Achievement and Assurance of Safety
Proceedings of the Third Safety-critical Systems Symposium, Brighton 1995
Redmill and Anderson (eds)
3-540-19922-5

Safety-critical Systems: The Convergence of High Tech and Human Factors
Proceedings of the Fourth Safety-critical Systems Symposium, Leeds 1996
Redmill and Anderson (eds)
3-540-76009-1

SAFECOMP '93
Proceedings of the 12th International Conference on Computer Safety, Reliability and
Security, Poznań-Kiekrz, Poland 1993
Górski (ed.)
3-540-19838-5

SAFECOMP '95
Proceedings of the 14th International Conference on Computer Safety, Reliability and
Security, Belgirate, Italy 1995
Rabe (ed.)
3-540-19962-4

SAFECOMP '96
Proceedings of the 15th International Conference on Computer Safety, Reliability and
Security, Vienna, Austria 1996
Schoitsch (ed.)
3-540-76070-9

Felix Redmill and Tom Anderson (Eds)

# Safer Systems

**Proceedings of the Fifth Safety-critical Systems
Symposium, Brighton 1997**

Springer

Felix Redmill

Felix Redmill
Redmill Consultancy
22 Onslow Gardens
London N10 3JU, UK

Tom Anderson
Centre for Software Reliability
University of Newcastle-upon-Tyne
Newcastle-upon-Tyne NE1 7RU, UK

ISBN-13:978-3-540-76134-1

British Library Cataloguing in Publication Data
Safety-Critical Systems Symposium (5th : 1997 : Brighton, England)
Safer systems : proceedings of the Fifth Safety-Critical Systems Symposium, 4-6 February 1997,
Brighton, UK
1.Industrial safety - Congresses 2.Automatic control - Safety measures - Congresses
I.Title II.Redmill, Felix, 1994- III.Anderson, T. (Thomas), 1947-
620.8 ' 6
ISBN-13:978-3-540-76134-1          e-ISBN-13:978-1-4471-0975-4
DOI:10.1007/978-1-4471-0975-4

Library of Congress Cataloging-in-Publication Data
A Catalog record for this book is available from the Library of Congress



Typesetting: Camera ready by editors
Printed by Athenæum Press Ltd., Gateshead, Tyne and Wear
34/3830-543210 Printed on acid-free paper

# PREFACE

The contributions to this book are the invited papers presented at the fifth annual Safety-critical Systems Symposium. They cover a broad spectrum of issues affecting safety, from a philosophical appraisal to technology transfer, from requirements analysis to assessment, from formal methods to artificial intelligence and psychological aspects. They touch on a number of industry sectors, but are restricted to none, for the essence of the event is the transfer of lessons and technologies between sectors. All address practical issues and offer useful information and advice.

Contributions from industrial authors provide evidence of both safety consciousness and safety professionalism in industry. Smith's on safety analysis in air traffic control and Rivett's on assessment in the automotive industry are informative on current practice; Frith's thoughtful paper on artificial intelligence in safety-critical systems reflects an understanding of questions which need to be resolved; Tomlinson's, Alvery's and Canning's papers report on collaborative projects, the first on results which emphasise the importance of human factors in system development, the second on the development and trial of a comprehensive tool set, and the third on experience in achieving technology transfer — something which is crucial to increasing safety.

Many contributions reflect technologies under development. Westerman and Ayton both report on work on the application of psychology to improving the understanding of the human component in safety-critical systems. Three papers from Bristol University demonstrate the breadth of research and the effort being invested in technologies which, it is hoped, will be available to industry in the near future. Gorski and Wardzinski's paper is complementary to Smith's in that it shows the means by which safety analysis is being made more comprehensive and, thus, more manageable by the practitioner. Whitty examines the difficult question of how to assess a product directly rather than having to rely on the indirect evidence acquired from an assessment of the development process. And Somerville demonstrates the practical application of state-of-the-art technology to the control of an excavator for use on construction sites. Loomes and Vinter interestingly point to how the nature of human reasoning can lead to errors in the interpretation of formal descriptions, despite the mathematical rigour of formal methods.

Rather more philosophically, Ahmad's examination of the 'language of safety' and Fenton's of the state of software engineering standards, particularly as they apply to the development of safety-critical systems, are significant and thought-provoking contributions which point the way to new lines of investigation.

The question running through the contributions to this book is, implicitly, 'Are systems getting safer?' More specifically, are we achieving safer design, construction, assessment, and operation? and, is our confidence (and, importantly, that of the public) in our systems increasing? Not only are these and other questions posed, but some answers are offered. Yet, the quest for safer systems continues: the questions of 'whether?' and 'how?' will be raised again at SSS '98 in February 1998.

FR and TA
October 1996

# CONTENTS

# The Safety-Critical Systems Club
sponsor and organiser
of the
## Safety-critical Systems Symposium


The Safety-critical Systems Symposium '97 is the fifth in the series of annual symposia organised by the Safety-Critical Systems Club. Its purpose is to raise awareness of safety issues and to facilitate technology transfer and collaboration among academics and industrialists.

Not only does the Club run this symposium; it also presents a number of 1- and 2-day seminars each year, and issues a regular newsletter (now in its 6th volume) three times annually.

The Club was inaugurated in 1991 under the sponsorship of the Department of Trade and Industry and the Science and Engineering Research Council (now the EPSRC) and is organised by the Centre for Software Reliability (CSR) at the University of Newcastle upon Tyne. It is a non-profit organisation, set up to provide a service to all, and it sets out to cooperate with all bodies involved or interested in safety-critical systems.

Since 1994 the Club has had to be self-sufficient, but it retains the active support of the DTI and EPSRC, as well as that of the Health and Safety Executive, the Institution of Electrical Engineers, and the British Computer Society.

The Club's purpose is to facilitate the transfer of information, technology, and current and emerging practices and standards. It seeks to involve both technical and managerial personnel within all sectors of the safety-critical community. By doing so, it can facilitate communication among researchers, the transfer of technology from researchers to users, feedback from users, and the communication of experience between users. It provides a meeting point for industry and academia and a forum for reporting the results of their collaborative projects. The goals of the Club's activities are more effective research, a more rapid and effective transfer and use of technology, the identification of best practice, the definition of requirements for education and training, and the dissemination of information.

Membership is open. Since the period of sponsorship expired it has been necessary to request an annual subscription — in order to cover planning, mailing, event organisation and other infrastructure costs — and this can be paid at the first meeting attended. Members pay a reduced fee for attendance at Club events and receive, free of charge, all issues of the newsletter, information on forthcoming events, and any special-interest mail shots. By participating in club activities, they also help to determine the Club's direction.

To join or enquire about the Club or its activities, please contact Mrs Joan Atkinson at: CSR, Bedson Building, University of Newcastle upon Tyne, NE1 7RU; Telephone: 0191 221 2222; Fax: 0191 222 7995; Email: csr@newcastle.ac.uk

# Safety-Critical Systems:
# Prescient, Presignifying, Public-Good Systems?


Khurshid Ahmad
AI Group
Dept. of Mathematical and Computing Sciences
University of Surrey
Guildford, SURREY
UNITED KINGDOM


## Abstract


The emergence of safety-critical systems and their much desired institutionalisation involves domain- and applications-independent issues that, in some sense, can be related to the interdependent issues of *knowledge*, *language*, and *ethics*. The design, maintenance, upgrading and decommissioning of existing networks for carrying, for example, water, gas or oil, are good examples where the triumvirate, that of knowledge, language and ethics, makes its presence felt. Knowledge based on experience, complemented by archives of regulatory, legislative, learned and popular texts, has to be articulated across and within groups of well-motivated individuals and organisations involved in making decisions regarding safe design, safe maintenance, safe upgrading and safe decommissioning in part, or in whole of an *in-situ* network. The safety-critical systems community should consider ways in which experiential knowledge of safe operation/design and so on can be collected and used in conjunction with a given textual archive through the use of an intelligent information system. This ever burgeoning knowledge, which is simultaneously being extended and being rationalised, is articulated through an expanded vocabulary but within a restricted syntax when compared with everyday language. The knowledge, and the language in which it is articulated, are both motivated by ways in which the individuals and the organisations *act*. This motivation can be related to what they consider they *ought* to do, what *obligations* and *duties* they have. The user-led SAFE-DIS project, concerned with the repair and design of urban water carrying networks, has demonstrated how the knowledge of safe design can be acquired, be formalised and be reasoned upon for autonomously generating hazard avoidance messages during the various design phases. The lessons learnt from this project are of relevance to the operators of equally complex energy networks, like electricity or gas networks, communication networks, and logistics networks.

# 1 How, what and why of a Safety-critical System

The developers of safety-critical systems attempt to build systems that might behave in a prescient manner on behalf of the users of the said systems. In other words, these systems should ideally have *foreknowledge* and *foresight* of potential hazards that may jeopardise the safety of a given user environment. The assumption of prescience posits the existence of an intuitively defined <u>knowledge of safety</u> which has a number of facets. One facet of this knowledge is the familiarity, awareness or understanding gained through experience (or study) of making systems and environments safe. Another facet of this knowledge deals with the states or facts of knowing that something is safe to use. Yet another facet is the sum or range of what has been perceived, discovered or learned about safety.

The knowledge of safety, one might claim, is collated, analysed and archived in various phases of the development of a safety-critical system. The developers of these systems use various methods and techniques such that the systems can presignify: A safety-critical system must be able to signify or intimate beforehand the existence of a hazard; once the hazard is signified or intimated, the system, in some cases, may then signify or intimate how the hazard can be avoided and how safe operation can subsequently be effected. The claim for presignification posits the existence of an intuitively defined <u>language of safety</u>. Such a language can be understood in a number of different senses. In one sense this language may be defined in terms of words and the methods of combining these words for identifying and avoiding hazards, and for preserving safety. A language of safety may also exist in the sense that the manner or style of expressing hazard- and safety-related information is distinct from the manner or style in which other kinds of information are expressed; this may involve the use of a distinct phraseology or terminology of safety, the use of a distinctive style in which safety-related information is composed. A language of safety might also exist in the sense that this is a *lingua franca* of the (health and) safety community including the operatives, end-users, and strategic thinkers involved in the operation, use and conception/modification of such systems respectively.

But what motivates the development of a system that is prescient and can presignify? What persuades a nuclear power plant operator to run the system safely? Why should an automotive manufacturer endeavour to use component-design software that produces a fail-safe design? How does a water company reconcile the cost of using expensive design software for designing water-carrying networks with its broader commercial objectives? Are safety-critical systems *pro bono publico*: public good motivated by some sense of duty or by some unarticulated moral code of conduct? A sense of obligation that motivates public good? Public good based on some idea of good/bad, right/wrong? Public good based on some notion of what *things* are good, right? Public good based on a conception of natural rights? The assumption that the specification and design of safety-critical systems is to a certain extent influenced by imperatives, like *ought, obligation, duty, right* and so on, and by judgements like the *desirable,* the *valuable,* the *good* , in itself suggests that there might be an <u>ethics of safety</u>.

The interaction between the imperatives (logicals) and value-judgements (axiologicals) in avoiding hazards is exemplified by the oft-encountered "No Smoking" sign - "No Smoking" is in effect a universal imperative that is effected, at a given point in time, by a singular imperative - "Do Not Smoke <u>NOW</u>!" The singular imperative entails the value judgement: "You Ought Not To Smoke". Indeed, one might generalise here and argue that a range of prohibiting icons displayed on plant and machinery, and many of the warnings generated by safety-critical systems, reflect this entailment relationship between imperatives and value judgements; the interaction between the logical and axiological contributes to the avoidance of hazard and the preservation of safety.

Safety-critical systems reflect the consensus of the interests of a number of stakeholders. One stakeholder is the vendor of a resource or service; the other stakeholder may be the end-user of the resource or service; people who are neither may also be regarded as stakeholders, e.g. regulatory bodies and public-interest lobbies. The safety of the immediate environment may be compromised by any of the stakeholders, by omission or by commission. Safety-critical systems, it appears, imply the existence of a contractarian moral theory that holds that an action, practice, law or social structure is morally permissible just in it, or principles to which it conforms, would be (or has been) agreed to by members of society under certain specified conditions. Such an approach to ethics is sympathetic to the neo-Darwinian ideals of 'market forces', 'perfect competition', 'trickle-down benevolence' and so forth.

This paper comprises an account of the recently completed SAFE-D*IS* project (Section 2) which has resulted in the development of an information system that can be used by novice engineers involved in the 'rehabilitation', that is repair and re-design, of urban water networks. Section 3 contains speculations about a 'language of safety' in terms of an idiosyncratic vocabulary and syntax used in safety-related communications. Section 4 attempts to introduce how studies in ethics can be related to the question of safety in general and to that of safety-critical systems in particular. Section 5 concludes the paper.

# 2 SAFE Design of Networks using Information Systems (SAFE-D*IS*) Project

This project was a three-year (1993-1996) collaborative venture, between a university (Surrey) and a vendor of specialist software systems (Wallingford), that dealt with safety related questions regarding the safe design, cost-effective repair and the subsequent hazard-free operation of large *in-situ* networks. These water-carrying networks serve large conurbations, comprise hundreds if not thousands of conduits (pipes) interconnected through an equal number of nodes (including inflows, outfalls, pumps, storage locations), and changes in the design and subsequent repair, sometimes termed *rehabilitation*, of such networks are classed as capital projects.

The SAFE-D*IS* project was joined by a group, the <u>SAFE-D*IS* Round Table</u>, comprising members from the private sector (three UK Water Companies), public sector (two local government-related organisations) and a UK civil engineering consultancy.

Knowledge related to the safe and cost-effective rehabilitation was acquired by the SAFE-D*IS* project team from human experts and from specialist texts. The text corpus comprises: safety guidelines and procedures, transcripts of expert interviews, learned papers and technical notes, legal texts including the complete Water Resources Act 1991 and a 450 page book that interprets the Act. The text corpus also comprises a terminology data base. All the texts relate in one way or another to the rehabilitation of water-carrying networks. This knowledge was structured in an information system for facilitating safe and hazard free rehabilitation of a part of the network[1]. The structured knowledge can be used to (a) help the experts examine their own knowledge, and (b) help novices to a greater or lesser degree throughout various phases of the complex rehabilitation process[2].

The SAFE-D*IS* project has identified five distinct groups of software systems that may help in the five key functions that are essential for the safe rehabilitation of complex networks (see Table 1). The integration of these systems was one of the achievements of the project:

Table1 - Overall functionality of Safe-DIS

| Function | Software Components |
| --- | --- |
| Access electronic documents | Full-text and hypertext management |
| Access rules and heuristics | Knowledge-base management |
| Modelling  complex network | Network simulation software |
| Sensitivity risk analyses | Risk analysis software |
| Model history and audit | Report generating systems |

One of the important decisions of the SAFE-DIS project was to use as much off-the-shelf software as was possible without compromising the high standards that are demanded for a safety-critical system. Thus the information system has access to proprietary simulation software, risk analysis software, text analysis software, knowledge engineering tools and data base management systems.

The SAFE-D*IS* project has also delved into the use of autonomous agents organised in a 'society of agents' through the use of constraint propagation. A prototype has been developed to show how autonomous agents can disseminate information about hazards to safety in a transportation network comprising a number of vehicles, drivers, freight types and transportation companies, see [Selvaratnam and Ahmad 1995].

The information system developed by the project team animates the behaviour of an experienced engineer setting a number of tasks for a less-experienced engineer to execute. This animation is based on an industry-wide rehabilitation procedure that involves over 20 specialist tasks distributed over 4 major phases (see Figure 1 below).

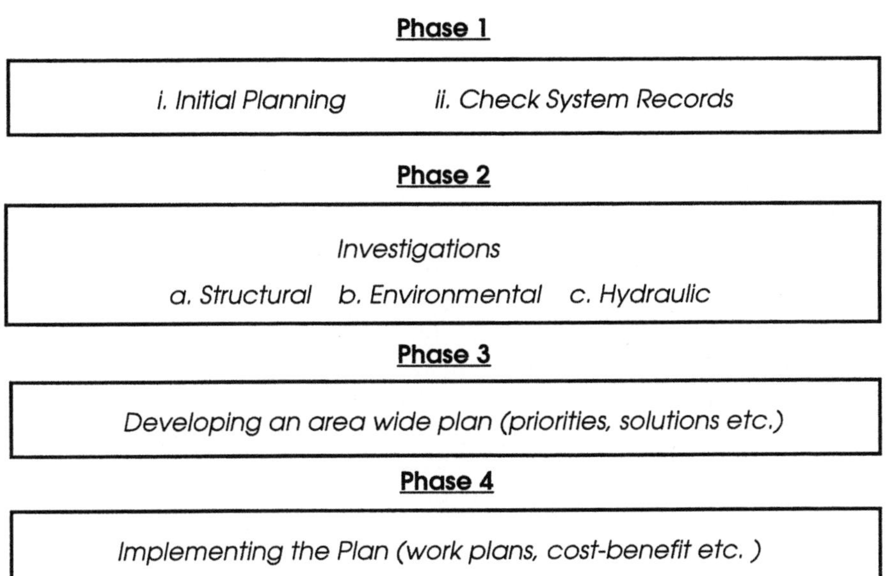


Figure 1. The Sewerage Network Rehabilitation Method (SRM) established by the UK Water Research Centre in consultation with the industry.

The first phase of systems' development in SAFE-DIS resulted in a conventional software system. Much like the conventional software system, including database systems or simulation engines, SAFE-DIS I relied on the user having sufficient motivation and/or knowledge to use any of the textual archives, the simulation engine, the propositions database or the automated procedures. Thus the system reacted to a knowledgeable user quite well, but for novices and, indeed, some experienced engineers, the operation of the system was somewhat baffling.

## 2.1 A Knowledge-rich, Integrated, Proactive Safety Information System

SAFE-DIS is a *proactive* system - a system which could execute the major and ancillary tasks outlined in each of the four major phases of sewerage rehabilitation planning  This proactive system acts in many ways like an expert system, wherein the system infers new facts from old, depending on the context, looks up and presents data from diverse sources, invokes other software systems and so forth.

This proactive system acts within the framework of the SRM method [WRC 1986].  During the execution of individual phases, and tasks within a phase, the proactive system provides **expert advice**, based on rules of thumb and other heuristics obtained from experts.  Proactively, the system can access excerpts and (optionally) full-text from a 'corpus' of texts, some of which are connected through hypertext links; a digital library that was built in close collaboration with the Round Table.  The advice is supplemented by access to **data bases** containing details of the various components of a given network and its geographical location, and supplemented by access to a industry-standard **simulation model**, namely *HydroWorks* developed and marketed by Wallingford Software.

Equally importantly the system keeps a **'diary'** of advice it gives to a user and the user can also enter his or her comment on the advice given.  **Risk analysis**, an important tool in the safety community, can be undertaken through the information system (SAFE-D*IS*) by the use of a low-cost, easy-to-use, and off-the-shelf system (namely Crystal Ball marketed by Decision Engineering Ltd.).  The information system also provides access to the **World-Wide Web** and through the Web provides access to up to date information related to engineering, legal and safety aspects of the aquatic environment as and when it becomes available on the Web. (See Table 2 for more details).  More advanced users of the information system have access to a **text analysis** system (namely *System Quirk*, see [Ahmad and Holmes-Higgin 95]).

Table 2. The functionality of the various components of SAFE-D*IS*. (The user interface of the Workbench is written in Visual Basic and runs on a PC.  The knowledge-bases are encoded in a variant of PROLOG.)

| Software Component | Narrative |
|---|---|
| Task Selection & Display | Enables an expert/manager to select tasks for a given project to be executed by a less experienced engineer. |
| Knowledge Management | Manages the knowledge base of the SAFE-D*IS* system and contains rules related to various rehabilitation tasks |
| Yellow Pages Management | Tracks the task a given user is executing and selects relevant excerpts (paragraphs and pages) from a full text-data base. |
| Safety Labels Management | Displays 'safety labels' during or after the execution of a rehabilitation task |
| Diary Management | Tracks when and how successfully each task was executed and notes it in a diary. |
| Report Generation | Generates an 'audit' report based on the contents of the 'diary' |
| Plug-In External Software | Helps to access data in proprietary data bases and acts as a front end for simulation software. |

## 2.2 Operational Details of the SAFE-DIS Workbench

The SAFE-*DIS* workbench offers two modes of operation: *professional* and *roster. The* 'professional' edition refers to a mode of operation designed for experts where they can either browse through the system, add more knowledge, modify or delete existing knowledge, and select some or all the phases, and tasks within the phases, for execution by less-experienced engineers. SAFE-D*IS*, thus, can be configured by senior design engineers in two important respects.  The first level of (re-)configuration is at the knowledge-levels whereby a designated user can add or delete subtasks to any of the four phases of the SRM method.  The second level of configuration is one where the senior engineer selects specific subtasks from one or all the four phases which he or she thinks should be investigated by one or more engineers reported to him or her.

The 'roster' edition refers the operation of the system by the novices where advice is provided and the novices can browse through the text corpus and access data bases and simulation models.  During the execution of each of the rehabilitation task, the user of the system, is guided through a question and answer session that includes display of *safety labels* on advice excerpts

During the interaction the workbench provides pro-active advice: excerpts of texts shown in the so-called 'Yellow Pages'.  Safety labels are sometimes displayed concurrently with the Yellow Pages. The labels come in three colours': *red* for mandatory warnings; *amber* for potential hazards; and *green* for safety notes. Access to full documentation, including Technical Notes (about 10 in number) authored by leading rehabilitation experts in the UK together with expert interviews and legislation is also provided by the workbench.

## 2.3 Knowledge Documentation: The role of the 'Round Table'

The project used a number of knowledge acquisition techniques reported in the artificial intelligence literature including face-to-face video taped interviews, structured walk-throughs, questionnaires, and interactive rule elicitation (see, for instance [Boose 1992], and references therein). Face-to-face interviews between experts and system builders were held on topics related to the safe rehabilitation of networks based on a case study. The questions in the interview were devised by the Round Table. Each interview was video-taped and the transcript of each interview was discussed by the Round Table during brainstorming sessions. The system builders extracted specialist terminology from the interviews, and extracted heuristics and rules. The transcript was marked up such that key parts of the interview could be extracted and linked to other documents through a hypertext browser.

The use of brainstorming techniques is seldom reported in the knowledge acquisition literature, yet the technique turned out to be very useful for devising questionnaires for the interviews, and subsequently for validating and verifying the acquired knowledge. In the SAFE-DIS Project, the brainstorming sessions were focused on the safety aspects of the specific phases of the rehabilitation procedures (cf. Figure 1). Individual members of the Round Table were given responsibility for providing knowledge related to given tasks in a specific phase; a detailed transcript of each of the sessions was prepared and circulated to the other members.

Corrections and modifications to the transcripts of the interviews and the brainstorming sessions were agreed by the Round Table as a whole. This consensus enabled the system builders to use verified and validated knowledge rather than the (unrevised) knowledge of a single expert as is the case in many knowledge-based systems' projects.

Structured walk-throughs helped in establishing the manner in which the various tasks within a phase are to be structured and in adding more knowledge for a task which the SAFE-D*IS* system could already execute.

Rule elicitation was used to develop automated/standardised procedures. These procedures, mini knowledge-bases, are particularly useful where the task is amenable to formal description, then automating according to a procedure agreed upon by experts will improve safety. During the structured walk-throughs the engineers provided rules and algorithms for various stages of the modelling process, e.g. choosing coefficients of discharge, accounting for unmodelled storage and checking for limits when doing catchment breakdown.

## 2.4 Simulation Engines

One of the subsidiary objectives of the SAFE-DIS project was to investigate the feasibility of intelligent front-ends and another was to investigate 'heuristics and rules of thumb' used in the development of simulation engines (see [Ahmad 95] for more details). One of the popular simulation engines, in the UK and abroad, is HydroWorks, developed by Wallingford Software. In the initial stages of the project it was thought that SAFE-D*IS* will essentially provide an intelligent front-end for HydroWorks: an intelligent system that will help in the selection of data for the simulation engine and in the interpretation of the output produced by the engine.

HydroWorks, like other simulation engines, appears to be adapting and/or incorporating a number of data management features, including improved file handling and data visualisation, and appears to have a better control of software releases. The vendors of HydroWorks are also taking onboard notions like quality management of the modelling process itself, including audit trailing and the generation of reports. This implicit development of an intelligent front-end, undertaken by the vendor, is a welcome development and helped the SAFE-D*IS* project team in focusing on safety-related aspects of the modelling process itself.

## 2.5 Dissemination of Results: The Workshop Series

The SAFE-D*IS* Project team held four workshops at the offices of the Round Table members, in the final year of the Project. Each of the day-long workshops comprised a presentation of the SAFE-D*IS* project and a demonstration of the system to audiences of company personnel ranging from new recruits to senior management. The presentations and demonstrations were followed by open sessions whereby attendees could come along and get a hands-on trial of the system and speak with the SAFE-D*IS* team. The day then closed with discussions which provided further feedback from potential end-users.

Each workshop was attended by over 20 attendees. By conducting the workshops during the life-time of the project it was possible to incorporate changes to the SAFE-D*IS* workbench. Indeed, these visits convinced the project team that what was required was a proactive system, rather than a reactive system, where a user is guided by domain-specific dialogue.

## 2.6 Failures, Hazards and Risks in Network Rehabilitation

Sewer network rehabilitation planning and the execution of such plans involves the presentation and analysis of a number of options. Some of these options require engineering judgements, whilst others may involve public health consideration, and yet others may include cost-benefit analysis. There is a significant emphasis in the rehabilitation planning exercises on considering regional priorities and, now there is an equivalent emphasis on 'setting [rehabilitation] priorities according to water company's business plan and investment objectives' in the UK. The Office of Water in the UK, a regulatory body with wide-ranging powers, sets rehabilitation targets for each of the ten water companies.

Although not explicitly mentioned the prevention of hazards and the anticipation of network failures, are amongst the principal considerations of the rehabilitation strategy produced by the UK Water Research Centre or the variations thereof created by water companies. Consider, for instance, the key term 'critical sewer' that is used very frequently throughout much of the rehabilitation documentation. The 'criticality' is defined in terms of 'sewers with most significant consequences in the event of structural failure'. A related term is 'core area': that part of a sewer network containing the critical sewers, and other sewers where hydraulic problems are likely to be most severe and require detailed definition within a flow simulation model. 'Acts of God', in their legal sense, also cause problems, so rehabilitation experts talk about 'catastrophic rainfall event', and event of return frequency far in excess of any sewerage design performance criteria, say, a 1 in 20 year storm. Sewer rehabilitation involves monetary expenditure and 'social costs'. The latter are defined as 'unclaimed business losses due to road closures, and the cost of extended journey times due to traffic diversions'.

Each of the four main phases of the rehabilitation of a sewerage network (Fig. 1) involves a number of considerations about the environmental impact of a rehabilitation scheme. Such considerations are elaborated in terms of 'systems failure', 'hazard prevention'. Tables 3a and 3b comprise the description of various tasks associated with two of the phases of rehabilitation planning. We have annotated these tasks with terms like 'failure', 'hazard' and 'precaution' to illustrate the implicit safety arguments therein:

## Table 3a. SRM Phase 1 Initial Planning and Records

| Phase 1.i | |
|---|---|
| **Task: Determine Performance Requirements** | |
| Hydraulic <u>performance</u>: (Failure) | Operational <u>performance</u>: (Failure) |
| Structural <u>integrity</u>: (Failure) | Environmental <u>Protection</u>: (Hazard) |
| **Task: Assess Current Performance** | |
| Use records of *flooding* (hazards) | |
| **Task: Is full investigation appropriate?** | |
| <u>Full</u> investigation (expensive), Abbr. Structural Investigation | |
| Abbr. Rural Investigation | |
| **Task: Check regional priorities** | |
| Investigate:i) Known causes (failure); | ii) areas of <u>imminent development</u> |
| (Precaution) | |
| iii)*poor storm sewage overflow* (Hazard) | iv) system with large number of <u>critical</u> |
| v) remaining <u>critical sewers</u> (failure) | <u>sewers</u> (failure) |
| **Phase1.ii** | |
| **Task: Check System Records** | |
| Depth of sewer; ground quality; marginally important traffic (failure) | |
| **Task: Identify <u>critical sewers</u>** | |
| Collect information (*Highly imp.* - roads) | |
| Apply screening proc. (sewer type A, B or C) | |
| **Task: Plan records upgrading and improving access** | |
| Produce Master Plan | |

## Table 3b. Phase IV. Implementing the Plan

| Task: Timing of Construction | Task: Maintain Hydraulic Model |
|---|---|
| OFWAT & Company Rehabilitation Targets (Failure) <u>Unit Cost</u> - criticality judgement | Audit trail must be kept for the model |
| **Task: Timing for Hydraulic work** | **Task: Review Drainage Area Plans** |
| Planned New Developments (Precaution) Legislation (Hazard) | Major changes New systems coming on-line OFWAT requirements |
| **Task: Design and Construction** | **Task: Deal with system failures** |
| Flooding (Failure & Hazard) | If a collapse occurs; |
| Operational Deficiencies (Failure) | • Make it safe |
| Structural Condition (Failure) | • Carry out repair |
| New Developments (Precaution) | • Monitor the area |
| Legislation Changes (New Hazards) Pollution (Hazard) | If a hydraulic problem occurs; • Develop solution • Record incident |
| External Influences (Failure) | • Implement solution |
| Risk | • Monitor solution |
| **Task: Monitor condition of critical sewers (Failure prediction)** | |
| Sewers must be cameraed and dated | |

# 3 Language of Safety?

This identification of the interaction between the knowledge of safe, cost-effective and hazard-free rehabilitation of a network, and the language used to articulate such knowledge leads naturally to the questions of moral imperatives and value judgements that, in our view, motivate the development of a safety-critical system. This is especially relevant in as vital a domain as water drainage that involves large capital expenditure: inadequate drainage can compromise human health and safety, and may cause the deterioration of the environment . This expenditure is initially paid for by a profit-oriented enterprise, like the water companies, for the greater good and prosperity of, for example, a conurbation living in a city or a borough. Safety-critical systems projects usually focus on one or two of the three sides of this knowledge-language-ethics triangle and it is not the usual custom to consider the dynamic between the key stakeholders and the information systems builders (see Figure 2).

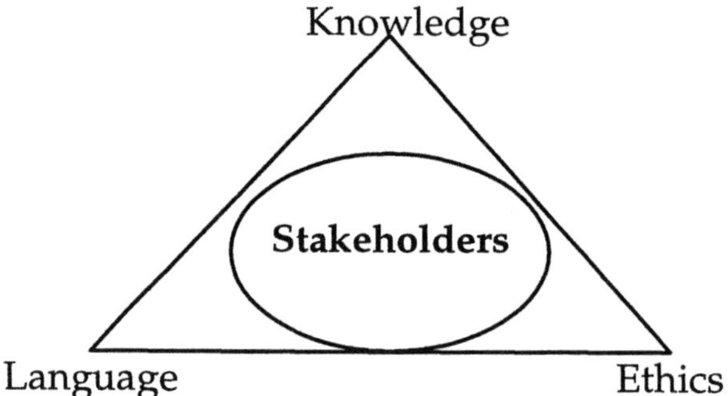


Figure 2. The safety triumvirate and the stakeholders.

Special language literature generally focuses on the language used by an identifiable community like scientists, engineers, artists and sportspersons, politicians and ideologues and so on. However, there are a number of instances where authors have discussed the use of language not just within one single discipline but across a number of disciplines. These para-disciplinary languages include 'language of negotiations', 'language of commerce', 'language of morals', 'language of ethics'. (The term 'para disciplinary languages was coined by Hoffman, [Hoffman 84]). To this list we would like to tentatively add the 'language of safety' - language used in preventing hazards and ensuring the safety of a system.

The emergence of subject-specific languages can be traced back to the development of a specialist subject, for example, through the emergence of neologisms, through novel syntactic constructs, interesting use of semantic relations, and pragmatically-adapted discourse patterns for addressing different (social) situations or *registers* as the linguists would like to call it; for example, expert-to-expert communication, expert-to-novice, expert-to-layperson and so on (see, for instance [Halliday and Martin 93]. The emergence of the para-disciplinary 'languages' is much harder to posit. It can be argued that such languages come into existence when there is a range of interests that come together in dealing with an artefact, an idea or an object. The emergence of the language of commerce can be attributed to the emergence of  vendors involved in trading (sophisticated) plant and machinery, in exotic edibles, or in intangibles like works of art, of fiction and so on, together with middle people (wholesalers, distributors) and consumers who were equally, if not more so, sophisticated.

The language of safety, languages of specialisms and language of everyday usage each can be compared and contrasted. The contrast between the languages shows that the 'languages' can sometimes be distinguished at two levels:  First, at the level of words or terms, the vocabulary level, and  second at the level of grammatical constructs used, the syntactic level [Ahmad and Salway 96].

## 3.1 The Vocabulary of Safety

The description of the various stages of a safety-critical system's development and operation involves the use of terms and phrases that are not *frequently* encountered in everyday language. We find terms like *risk, safety, tolerance, hazard, failure/fail* and so on, together with compounds like *risk classification, risk analysis, hazard identification, safety argumentation, safety integrity, safety property[3], tolerability of risk (TOR), 'as low as reasonably practicable' (ALARP), fail-safe mechanism, failure mode, functional safety.*

These terms are used in a number of papers in the safety-critical systems' literature across a number of domains. However, there are a number of terms and phrases that are used in some domains and not others, for example:

Table 4. Quasi domain-oriented safety terms

| Term | Domain | References |
|---|---|---|
| accidental radiation overdose | Radiation Physics | [Thomas 94: 35] |
| contingent valuation question own 'statistical' life. | Transportation Systems | [Jones-Lee and Loome 95: 21] |
| consequent led analysis of safety and hazards (CLASH); | Advanced Robotics | [Seward et al 95:166] |
| sneak circuit analysis | | [Elliot et al 95: 147] |
| beyond design basis accidents; incredibility of failure of key pressure vessels. | Nuclear Engineering | [Hughes et al. 95: 171-187] |
| Property damage type failure | Storm sewer design | [Yen 75] |

The quasi domain-oriented terms are used in conjunction with the terms founds across the domain (mentioned above). The proliferation of safety-related terms and phrases, both domain independent and quasi-domain dependent terms and phrases, in safety-critical systems literature, is mirrored in safety-systems literature in general. The important point to note here, and as we subsequently show also, is that the safety terminology and phraseology is neither used in everyday language nor in the domain-specific literature.

In order to understand the vocabulary level difference between safety texts and domain specific texts on the one hand, and the safety texts and general language texts on the other, we followed the following methodology. First, we created a corpus of domain-specific texts from a specialist domain and a corresponding corpus of texts dealing specifically with safety in that domain. Second, we created a frequency ordered word list for <u>each</u> of the two corpora. Thirdly, the first 100 most frequent words in each corpus were selected and compared. Having completed the comparison between specialist texts and specialism-related safety texts, say Radiation Physics and Radiation Safety texts (or Sewer Design and Sewer Safety texts), select a representative and contemporary corpus of general language texts, find the 100 most frequent words and compare this list with specialism list and 'safety' list. We used the 20 million word Longman Contemporary Corpus of English language that contains everyday language as used in (quality) newspapers, magazines, works of fiction, belle lettres and popular science texts (see [Aijmer and Altenberg 91] for more details of the Longman Corpus).

The radiation physics texts (55 in all comprising a total of over 85,000 words) and radiation safety texts (21 texts comprising about 190,000 words) were collected through the World-Wide Web using various search engines. The radiation physics texts include learned papers, advertisement for conferences and courses, popular science texts in radiation physics and radiotherapy. The radiation safety texts comprised research papers, safety manuals, public awareness documents from the Paris-based Nuclear Energy Agency and the UN-World Health Organisation, reports on nuclear accidents like Chernobyl, official documents from the US-regulator, the Environmental Protection Agency, and advertisements for goods and services in radiation safety.

The results of the comparison between the three genres, Radiation Physics (specialism), Radiation Safety and everyday language (Longman Corpus), are shown in Table 5. In order to save space, we show the aggregated frequencies of batch of ten words. Note that there is only <u>one</u> noun amongst the 100 most frequent words in everyday language (Table 5, column 1) and even that is found in the lower frequency regions; the 100 most frequent words comprise just under 50% of all the words found in the Longman Corpus. However, the nouns make up around 40% of the 100 most frequent words in the specialist corpus and in the safety corpus (all nouns in the 100 most frequent list in Table 5 have been underlined).; 100 most frequent words make up just over 40% of all the words found in the two corpora. (The $3^{rd}$, $5^{th}$ and $7^{th}$ columns contain the values of relative frequency which is equals to the absolute frequency divided by the total number of words or tokens).

The Radiation Safety corpus contains terms like *accident hazardous, contamination, incident, protection, contamination, and fallout.* All these words were also present in the Longman Corpus, but in the Radiation Safety Corpus these words are used at least 30 times more frequently and words like 'fallout' over 2000 times more frequently.

Table 5. The 1st percentile of a frequency-ordered list of words in the Longman Corpus compared with the Radiation Physics and the Radiation Safety Corpora.

| Rank | Longman/ Lancaster Corpus 20 million token | Rel. Freq (%) | Radiation Safety Corpus 189169 tokens | Rel. Freq (%) | Radiation Physics Corpus 85109 tokens | Rel. Freq (%) |
|---|---|---|---|---|---|---|
| 1-10 | the, of, and, to, a, in, it, that, i, was | 22.54 | the, of, and, in, to, a, for, or, is, nuclear | 24.28 | the, of, and, in, to, a, for, is, are, with | 22.41 |
| 11-20 | he, is, for, as, with, his, on, you, had, be | 7.91 | be, this, by, as, on, from, with, that, are, at | 6.01 | be, on, nuclear, radiation, from, by, at, as, data, that | 5.56 |
| 21-30 | not, she, they, her, by, this, from, or, have, are | 4.99 | which, an, bq, shall, not, may, chernobyl, 137cs, was, other | 3.42 | this, energy, dose, mev, or, an, which, neutron, cross, it | 3.45 |
| 31-40 | which, we, all, were, an, one, there, said, him, so | 3.69 | accident, such, waste, radiation, radioactive, kg, safety, it, any, have | 2.50 | protection, image, was, have, beam, used, these, measurements, we, can | 2.35 |
| 41-50 | what, would, their, when, if, no, my, been, out, up | 2.71 | report, data, health, section, will, power, has, no, level, radioactivity | 2.06 | will, been, research, sections, has, al., et, electron, also, beams | 1.85 |
| 51-60 | them, more, about, can, me, who, like, into, has, then, | 2.27 | than, all, article, site, convention, emergency, fuel, i, these, time | 1.73 | physics, were, not, may, other, power, new, more, results, absorbed | 1.64 |
| 61-70 | could, do, will, time, only, some, other, its, than, now | 1.82 | were, fallout, one, levels, information, reactor, been, deposition, protection, pci | 1.54 | system, there, measurement, dosimetry, use, reactor, university, high, imaging, treatment | 1.46 |
| 71-80 | two, very, these, over, any, did, down, way, back, first | 1.40 | more, if, contamination, hazardous, available, energy, x, also, their, national | 1.40 | such, than, our, technology, total, well, using, all, figure, accelerator | 1.30 |
| 81-90 | man, know, just, see, may, our, how, even, well, your | 1.18 | its, per, used, environmental, following, research, after, but, Maine, exposure | 1.25 | about, clinical, section, medical, gamma, medicine, science, some, ray, fission | 1.19 |
| 91-100 | such, where, because, after, much, made, before, little, most, through | 1.08 | contracting, system, paragraph, public, high, means, response, incident, substances, plutonium | 1.14 | radiotherapy, up, but, conference, exposure, calculations, electrons, studies, low, one | 1.08 |
| | TOTAL | 49.59 | TOTAL | 45.34 | TOTAL | 42.28 |

The Radiation Physics corpus emphasises the more positive aspect of the subject and amongst the frequent nouns in the 100 most frequent list include *energy, Mev[4], neutron, dosimetry, image, beams.* The nouns used in the Radiation Physics are used with much higher frequency than in the Longman, some like 'electron', 'image' and 'beams' over 100 times more frequently; words like 'neutron' and 'reactor' over 1000 times more frequently; and indeed contain words that are not found in Longman Corpus at all.

Similar results were obtained when we compared texts in Sewer Design (53 texts made up of over 140,000 words) with Sewer Safety texts (22 texts containing over 40,000 words) and contrasted the results with the Longman Corpus. We are adding more texts to our respective corpora such that each of the corpus contains around 250,000 words of texts for us to make more definitive claims. However, we see that the vocabulary used in safety documents has characteristics that places it in between general language texts and the specialism's texts. Our preliminary analysis of morphological analysis and compound words shows that this is indeed the case; safety texts appear to have their own idiosyncratic signature.

## 3.2 The Syntactic Characteristics of a Language of Safety

As our knowledge of safety and safety-critical systems grows, the vocabulary associated with such knowledge grows quite dramatically with words and terms that are deemed to be outdated falling by the wayside. But what of the grammar that is used to put those English, Chinese or Swahili vocabulary items into, say, English, Chinese or Swahili sentences? The oft-repeated assertion in linguistics (see, for example, [Halliday and Martin 93]) and in the philosophy of science [Gerr 42] is that this expanded vocabulary is accompanied by a restricted syntax: 'the rationale of linguistic formulation as a whole through progressive reduction of syntactic complexity to the absolute minimum established by the requirements of formal logic analysis and exposition, as well as through extended use of functional terms' [Sager, Dungworth and McDonald 81: 185], citing [Gerr 42].

Given that the safety-critical systems literature is written in accordance with its strong information intention, a large number of the sentences in the literature should be of a declarative nature. The structure of such sentences is dependent on the grammatical relationship between the nominal groups it contains. An equally important intention of those who write about safety-critical systems, particularly those involved in supervising the operation of such systems, is to instruct (operators) about the nature and function of a plant, machinery and so on. These instructions are encoded as imperative sentences: concise sentences usually prefixed with numbers or letters. The other sentence types like interrogative and exclamative sentences are conspicuous by their absence in the safety-critical literature.

<u>Declarative sentences: conditional sentences for expressing causality.</u> The dependence of consequents on antecedents is expressed by generally having the *consequent* in the main clause and the antecedents in the *if* clause. Consider, for instance:

**If tolerance limits are set narrower than the natural spread, the manufacture of defective parts is inevitable.**

**If the total sulphur content (as $SO_3$) is greater than the acid soluble sulphate content and if a significant quantity of acid soluble calcium is present, then the shale should be regarded as potentially troublesome and an inspection of the quarry or tip carried out.**

Sometimes *if* clauses are replaced by a small number of compound prepositions and conjunctions, like *in case of, in the event of, on condition that, so long as, providing that, provided that,* e.g.

**In case of fire, all workers will leave the building**
**In this form, the system is fail-safe in the event of possible disconnection.**

At other times the *if*-clauses are replaced by participle clauses:

**Given sufficient turbulence in the combustion chamber, detonation is likely to occur.**

<u>Declarative sentences: The attributive use of connection verbs like *be, have,* or *give.*</u> Consider the following elaboration of the term safety integrity due to [Bell 1994]. He argues that safety integrity [...] must be of such a level as to ensure that

**The *failure frequency* of the *safety-related system* <u>does not exceed</u> that required to meet the *tolerable risk-level***

The above sentence shows how three safety-related terms, failure frequency, safety-related systems, and tolerable-risk level, are interrelated. The first two are related in a simple structure comprising two nominals linked by verbs such as *be, have* or *give*. The first [and second] are then related to the third term through a more complex semantic process, that of *grading*, which is normally accomplished through the use of adjectives and adverbs: except here the verbal phrase does not *exceed* is used to express comparison in relation to a lower degree.

<u>Declarative sentences: Grammatical categories and causality expression</u> [Brazendale and Jeffs 94] have elaborated the 'Safety Lifecycle' concepts with the help of a number of control-system failure case-studies. They take 'a project-oriented view of the safety lifecycle is used to highlight when the mistakes were made and what precautions are needed to prevent it

happening again' [Brazendale and Jeffs 94:67)]  A number of declarative sentences expressing cause and effect, through the use of verbs, conjuncts, preposition, or the +*ing* form of the verb, show the cause of failure (see Table 6a).

Table 6a. Some declarative sentences used in safety-critical literature

| Lifecycle Stage | Domain | Failure cause/effect | Syntacto/Semantics |
|---|---|---|---|
| Error/ Inadequacy Specification Error | Automated Transit System | The stop button on the controller had the effect of causing all the controller outputs to revert to power up states | Subject ('stop button on the controller') denotes the cause and predicate the effect |
| Inadequate Specification of safety integrity | Microwave Cookers | When the interlock and sensor conditions were satisfied a contactor was operated, and the microwave power switched on. | Use of main clause ('a contactor ...' + adv, clause) to show effect and the co-ordinated clause to show effect. |
| Inadequate design and implementation | Chemical Spillage | In this incident, the operator inadvertently 'called up' the schedule of equipment [...] on which he was not working, and consequently the wrong value was operated. | The adjunct *consequently* was used to show cause and effect relationship between two parts of the sentence.  One could also use *hence, so, therefore* type of conjuncts |
| Inadequate operation and maintenance | Hydraulic operated guillotine | Interruption of the light curtain [...] caused the *blade* to move downwards, instead of upwards to its safe position. | Subject (interruption of the light curtain) denotes the cause and the predicate the effect. |

<u>Imperative sentences</u>  Such sentences occur more predominantly in the language of safety than, say, in the language of science or in the general language of everyday.  These sentences, often paragraph long, appear to be concise and usually each clause in the sentence is labelled by a number or a letter.  The following two examples show two types of imperative, safety-related sentences: one for operating plant and machinery, and the other used in the description of the duties of a safety operative:

Table 6b. A selection of imperative sentences used in safety literature

| Plant/Machinery | Quality Assurance |
|---|---|
| Domestic power drill: To change speed -<br>1.  Ensure motor is stationary<br>2.  Life the shift lever..................<br>3.  Rotate half turn<br>4.  Snap back to closed position<br>5.  Rotate chuck half a turn by hand...<br>[Sager, Dungworth & McDonald 80] | Responsibility of a QA Manager:<br>(i) Ensure that the project is fully defined<br>(ii) Ensure that the staff employed are...<br>(iii) Engender a positive attitude towards safety<br>(iv)  Ensure that quality auditing is done correctly<br>[Kirk 94: 82] |

## 3.3 The Uses of a Language of Safety?

The articulate individuals involved in the design, implementation and operation of safety-critical systems, are trained to be precise in their speech and their writing. It can, perhaps, be argued that the range of safety-critical systems perform as they are expected to, that is delivering an uninterrupted service free from serious failures, is in some measure due to this precision in the language of safety: expanded vocabulary and restricted syntax is one way of achieving the precision. One corollary of our observation will be to argue that serious failures of such systems may be caused by a breakdown in linguistic communication between those charged with safety and the public at large.

A systematic and objective study of how safety arguments are prepared and communicated can be undertaken by looking at the linguistic strategies used in such argumentation.

# 4 Ethics of Safety

The study of the knowledge of safety and the language of safety for a specific domain tells us that people involved in a safety-critical system may subscribe to different value systems. Such variance can sometimes be perceived by their actions. For philosophers, like Hare, who are interested in ethics 'actions are in a peculiar way revelatory of moral principles' [Hare 90:1]. Hare, according to [Williams 93] is a *utilitarian* philosopher, who claims that moral judgements are prescriptive (loosely it amounts to saying let-so-and-so-be-done) and they are universal. For another group of philosophers, the so-called *contractural* or *contractarian* philosophers, 'an act is wrong if its performance under the circumstances be disallowed by any systems of rules for the general regulation of behaviour ...' (T. M. Scanlon cited in [Williams 93:75]). Contractarian moral theories have been used in a range of interesting domains ranging from market economics, social justice to nuclear disarmament/rearmament (Gauthier and his colleagues in [Vallentyne 91] are good examples of contractarian moral philosophers). 'Value theory', see for instance [Moore 73] also can help in understanding a value system. A value system is a pre-requisite for any culture; those interested in the establishment of a safety culture should also think of a value-system.

The moral values of the key players in major environmentally sensitive enterprises, including those involved in making the so-called 'safety arguments' (see below for details), are couched in an ethical terminology and constructs that are at best ambiguous and at worst misleading: ethically-fuzzy words like 'promise' and 'courage' were used by one set of protagonists and they are countered by equally muddled terminology comprising words like 'treachery'. Williams has noted that these words are confusing because they seem to express 'a union of facts and values' [Williams 93:129]. It would not do, as Williams reminds us, for some moral philosophers, like [Singer 93], 'to increase a sense of indeterminate guilt in their reader', because such a line of argument does not pay much attention to 'the theoretical basis of the argument' and is likely to be 'counterproductive' [Singer 93:212] For Hare 'confusion about our moral language leads, not merely to theoretical muddles, but to needless practical perplexities' [Hare 90:2]. The privatisation of water industry in the UK and the debate between the water industry operatives, their regulator, the pressure groups, and occasionally members of the public, have led to 'muddled' and 'counterproductive' arguments at one time or another.

No matter which of the ethical styles one chooses, utilitarian, contractarian, value-theoretic, the key point is that one should adopt one especially in the context of safety-critical systems. In safety-critical systems, the emphasis is on logic - the welcome wide-spread use of formal methods is a sign of this; on the physics/chemistry/biology of such systems; and, although most people miss out the third part of the Kantian trilogy, on ethics. The use of ethnographic techniques is some compensation for not looking at moral principles or for leaving them unsaid (see below).

We present an analysis of three case studies from the safety-critical literature to show the implicit nature of moral issues in the literature before we briefly look at the origins of computer ethics and at the advice the UK IEE has for its members involved in safety-critical systems. We then examine two styles of ethical theory: Hares' analytical model and Gauthier's contractarian model. These styles may help to articulate the moral issues implicit in safety-critical dialogues.

## 4.1 Visualising Values? Three Case Studies

### 4.1.1 Illustrating the Safety Case (Safety Argument): Diagrams and Formal Networks

The safety argument is 'an informal argument, embodying engineering judgement rather than a strict formal logical argument' [IEE 92: 6], and is required in many industrial sectors by their relevant regulatory authority before a system can be put into service. The safety argument is prepared by designers and vendors in conjunction with the end-user purchasers for two reasons: First, to demonstrate the confidence the system designers/vendors and purchasers have in the safety of the system; second, to demonstrate that 'even though an unforeseen event may occur, nevertheless all reasonably determinable safety-related concerns were considered, and dealt with properly. This may provide an important legal defence' [ibid.: 6]. For IEE, a safety argument is 'a good engineering practice'.

The preparation of a safety argument is in part motivated by the presence of a regulatory body, partly by the needs of the designers and the vendors to sell such systems, and partly due to the needs of the purchaser and the potential end-user. The safety-argument also, perhaps more importantly, can or should be used for defending the designers and the purchasers if the system fails and endangers life, and/or property. The 'engineering judgement' embodied in the argument is complementary to 'strict formal logical argument'; perhaps the judgement is based in part in domain knowledge, in part in the experiential knowledge of operating safe plant and machinery, and, for us, in part in the value system espoused by the designer/vendor and the purchaser of the system. But more of the 'value system' later.

Safety argument is also used synonymously with terms like 'safety case', 'safety assessment report' and 'safety justification'. For academics committed to the study of safety-critical systems, a safety case is defined as a 'collection of data drawn from multiple sources, representing diverse disciplines, which must be assessed by a range of technical and non-technical experts, who exhibit a range of interests'. [Gurr 95] has advocated the use of formal methods in the design and modelling of safety-critical systems in that such systems can help in 'removing ambiguities in [system] specifications and [in] making explicit assumptions which might otherwise remain hidden' [Gurr 95: 385]. The author argues that because people of different backgrounds and interests contribute to a safety case, the complex information which involve formal methods be presented through the agency of 'a well founded diagrammatic representation'. Such an approach, Gurr thinks, when suitably augmented with automated support, will help in modelling concurrent systems; continuing expressive powers (of an algebraic system like calculus of communicating systems) with the usability made possible, for example, through statecharts by 'both experts and non-experts' [Gurr 95: 395].

This diagrammatic presentation of a safety argument, underpinned by logico-algebraic proof of systems, takes us a step closer in understanding the motivation of the various parties involved in the argument. Furthermore, Gurr's work shows how the confluence of 'engineering judgement' and 'strict formal logical argument' can be visualised. The exploration of such techniques by nuclear power plant operators and others is a welcome step indeed.

### 4.1.2 Observing and Describing Requirements: The case of the Situated, Reasonable 'Observer'

[Bentley et al. 92a] have discussed the use of methods and techniques of ethnography in the design of control systems, particularly in air traffic control systems; Bentley et al. have argued that the contribution of sociology and anthropology is through the description and analysis of the 'real world' in which the safety-critical system will be implemented and operated.

Bentley et al. claim that 'the roots of ethnography are in anthropology so ethnographers are trained to avoid making judgements about a social situation or process and, as far as possible, to avoid letting their own prejudices interfere with their observations'. This claim becomes even more interesting when by way of elaboration the authors observe that '[B]y contrast, engineers *must* make judgements, as to what is and is not significant' [Bentley et al. 92a: 5]. The sociologist or the ethnographer (these terms are used interchangeably) is interested in social structures and the dynamics between the people the structures support. Whether or not the sociologists avoid their own prejudices and act as rational agents is not a question in which we are interested here. What interests us are four observations: (a) the observations about the various power relationships Bentley et al. have noted, for instance, between the air-traffic controller and other members or his or her team; (b) the various observations about alternating certain manual tasks; (c) Air Traffic Control (ATC) is not as tightly circumscribed and as highly role governed as it might appear when one reads ATC manuals: 'work-in-hand' is a key determiner, rather than any formal procedure, in the control room and there is a 'highly informal "working division of labour"' [Bentley et al. 92b] and (d) the various observations comprising emotive terms like 'important'/'unimportant', 'idle'/'useful chat' and so forth that takes place in ATC. Clearly, what the ethnographer is doing is to note and describe the value system of the observed and consequently evokes notions of 'duties', 'obligations', 'rights'. This evocation is presented as constituent <u>facts</u> of the social structure of, say, the air traffic control room.

There is no doubt that ethnography is becoming a popular framework for identifying, observing, documenting and analysing culture (patterned beliefs and behaviour) in communities, institutions and target populations under difficult field circumstances [Weeks and Schensul 93: 53]. [Fetterman 93] has claimed ethnographic methods used in policy planning and implementation help to translate knowledge into action. The author discusses specific strategies for addressing adversarial audiences, which include listening to others, stress on qualitative/descriptive research rather than qualitative/analytical; 'these strategies involve communications, collaboration and advocacy; the utility of these strategies depend crucially on how ethnographers use language in a range of different disciplines' [Fetterman 93b: 164].

Ethnographic methods and techniques bring us closer to studying the power-related aspects of the value of systems of the operators of safety-critical systems, what Fetterman calls 'patterned beliefs and behaviour'.

### 4.1.3 Viewpoints in Safety Requirements: Discovery and Analysis

Some authors argue that conventional requirements analysis is not suitable for safety-critical systems in that safety is a whole-system consideration and not just restricted to software systems, thus systems cannot be discussed in isolation. Furthermore, conventional systems do not focus on 'requirements discovery' due to their preoccupation with analysis, and consequently cannot be very helpful in 'discovering what the system *must not do* rather than what services it should provide'.

[Kotonya and Sommerville 94] and [Seward et al. 95] talk about organising and structuring requirements of a system; clients, especially how he or she will receive services from the system and send control information and data to the system. In addition to these 'direct' clients, there is a need to organise and structure the requirements of those who may have an 'interest' in some or all of the services but do not send control information and data to the system; the so-called 'indirect' clients, which include transaction security, enterprise evolution and automation and its effect on extant (manual) labour. The authors have introduced the notion of 'direct' and 'indirect' viewpoints so as to make explicit the relationship between functional and non-functional requirements.

The authors use a 'set of user definable severity and risk schemes based on the UK Ministry of Defence Standard 00-56' [Seward et al. 95: 162], and use 'fault tree analysis' for analysing hazards wherein for each identified hazard '...a fault tree is produced which traces back to al possible situations which might cause the hazard' [Kotonya and Sommerville 94: 13].

Terms like hazard analysis and risk classification have ethical connotations: Hazard analysis involves a description of all those with a viewpoint - the operators, their supervisors and manager, safety officers and so for, and helps the requirements engineer to identify hazards, to analyse hazards and to note hazard event information. Risk classification involves the description of 'severity schemes' that range from catastrophic to fatal and also from 'severe' and 'minor', and the description of probability category of the scheme which ranges from frequent, probably, occasional, remote, improbable and to implausible. A scheme in the Defence Standard parlance involves a plan where 'the risks associated with a system failure can be classified with a view to deciding on whether or not these risks are acceptable' [94: 14]. The authors also discuss 'conflict analysis', after safety analysis, wherein various viewpoints are synthesised to produce 'negotiated changes' in the requirements description.

The *viewpoints* perspective of a safety argument is amongst the most ambitious of our three case studies especially by its emphasis on 'whole system considerations'. Sommerville and his associates are striving to articulate complex issues in moral philosophy as can be seen in their exemplar risk classification schema for instance.

## 4.2 Safety in an Ethical Context

The three case studies show interesting uses of the terms 'safety' and 'risk'. The textual examples highlighted indicate the responsibility of the engineer for avoiding hazards, minimising risks, ensuring safety of operation and so forth.

Our contention is that any safety argument is motivated by the value systems of those who build, operate and use such systems. The moral principles and the value systems of the three parties - computing professionals, (network) operators and the end-users - may or may not be congruent. This potential lack of congruence can be superficially attributed to the 'arrogance' of the safety-system builders, in that they are concerned mainly with machine-related, technological issues; attributed to the 'greed' of the system operators, in that they are concerned with maximising their profits; and the 'ignorance' of the end-users in that for many reasons they cannot appreciate technological issues faced by the system builders, and for reasons of envy, fail to understand the business strategies of the system operators. One can argue that unless the moral principles and value systems are clearly articulated, one will have this lack of congruence. A clarification of the meaning of the terms 'safety' and 'risk' in an ethical context is perhaps in order here.

The British Standards define safety as 'freedom from unacceptable risk of personal harm. The likelihood that a system does not lead to a state in which human life or environment are endangered' (BS0: pt3, clause 10.11). The UK IEE defines a safety-related system as 'a system by which the overall safety of a process, machinery or equipment is achieved'. This definition is elaborated by a typology of safety related systems that includes two types: First, the class of systems in which separation of control and safety functions cannot be made, like flight control system. Second, those systems that are designed to respond to hazardous conditions, and protect against them, independently of other control systems.

The term risk is used in the safety literature and in general language much in the way of its Italian root *risco, rischio* which in turn are formed from *rischiare* meaning *run into danger*. The modern sense of the term includes (a) Hazard, danger; exposure to mischance or peril; (b) the chance of hazard of commercial loss (cf. *The Shorter Oxford English Dictionary*). *Risk assessment*, a term used frequently in safety-critical systems literature, is defined in specialist dictionaries as 'a description of the safety of a plant in terms of the frequency and consequence of any possible accident' [Walker 95: 943]. The UK Institution of Electrical Engineers in their 'Professional Brief for the Engineer' on safety related systems defines risk in the parlance of risk assessment : Risk is the 'likelihood of a specified hazardous event occurring within a specified period or in specified circumstances'. And the purpose, objective or task of a safety related system is to 'ensure that the risk is reduced to - at worst - this [pre-defined] tolerable level, and attention must be paid to any legal requirements for the particular case in question' [IEE 92:6], articles 1.25, 1.27 and 1.29.

The legal requirements are outlined in a number of regulatory Acts of the UK Parliament. For example, the Health and Safety Act 1974 and the Consumer Protection Act 1987, place the burden of being reasonable on the employers and vendors. The adjective *reasonable* used in the citations from various Acts of the UK Parliament discussed above can be interpreted in different, perhaps overlapping, senses as: (a) sensible or sane; (b) requiring the use of reason; (c) rational by being not 'irrational'; (d) moderate by being not extravagant; (e) suitably costed; (f) of such an amount, or size, number, etc., so as to appear appropriate or suitable. 'Sensible', 'moderate', 'rational', 'appropriate' are the sentiments also invoked by IEE's definition of reasonable: the word 'reasonable' is used here to stress the fact that 'the search for safety is often a trade-off between safety-assurance and time, effort and money spent on acquiring that assurance'.
The responsibilities of the engineers and managers are further elaborated by a code of practice which demands that the engineer or the manager should [IEE 92 : 21, §5.8)[5]]:

(i) at all times take <u>reasonable care</u> to ensure that there are no unacceptable risks to safety;
 (ii) <u>not make claims</u> for their work which are untrue or misleading...;
(iii) <u>accept personal responsibility</u> for all work carried out by them or those reporting to them
(iv) <u>take all reasonable steps</u> to keep abreast of new developments in relevant aspects of science and technology;
(v) <u>declare their limitations</u> regarding areas in which they are not competent;
( vi) <u>take all reasonable steps</u> for disseminating their knowledge about risks to their managers and those to whom they have a <u>duty of care;</u>
(vii) <u>take all reasonable steps</u> to ensure that those working under their supervision [...] are competent and know their responsibilities.


The code of practice for engineers and managers discussed above prescribes (items *i, ii, iv, vi*) and describes (items *iii, v and vii*) the conduct of an engineer manager involved with safety-related systems. [Whitbeck 95], who has discussed similar issues in some considerable detail regarding an engineer's or chemist's responsibility for safety, calls such issues, issues of <u>professional ethics</u>.

The descriptive and prescriptive statements (i - vii above) can be viewed from three perspectives. First, they can be treated as a kind of statement of intent, a declaration of interest and commitment by the computer professionals. Second, the statements, perhaps more importantly, can be viewed as statements about how computer professionals *ought* to act in general: what they *ought* to do, what are their *duties*? How can their *obligations* be defined? Which of their *acts* can be deemed right, and which *wrong?* Third, for some, an equally important viewpoint is that the statements could form the basis of the value system for safety-related professionals wherein they could discuss what is valuable for them (and society by extension), what is desirable, and what is good. This can, perhaps, form the basis of a culture of safety.

The second perspective above, with its emphasis on ethical terms like *ought, obligation, duty, right, wrong* is sometimes perceived as the domain of *moral theories, moral principles* or simply morals. *Deontologists*, who espouse moral theories, like Immanuel Kant argue that duty is prior to value and some of our duties, like promise-keeping, are independent of values. The third perspective is often discussed under the rubric of axiology, the study of value in general. Teleologists argue that our only duties have reference to ends and to produce value, or perhaps to distribute it in certain ways.

## 4.3 The Origins of 'Computer Ethics'

Abbe Mowshowitz was amongst the first scholars who talked about the 'morality' or 'value' of information processing in human affairs and noted the post-Second World War computing systems serve one of two general social functions: the co-ordination of diversity or the control of disorder [Mowshowitz 76]. The discussion in the literature on computer ethics still focuses on the issues of co-ordination and control. Safety and safety-related systems do not, as a rule, figure prominently in this literature. In her very important contribution to computer ethics, Deborah Johnson [Johnson 94] argues that 'the bottom line is that all of us will benefit from a world in which computer professionals take responsibility for the computing in our society, at least, when it comes to safety, reliability, and security, but also for other effects' [Johnson 94 : 55]. However, issues related to safety, and safety-related systems - terms which do not appear in Johnson's index - are discussed implicitly.

[Mitcham 95] has reported that the influential political think-tank, the Brookings Institute, together with IBM and the Washington Theological Consortium, have sponsored the Computer Ethics Institute. This Institute has adopted and promotes a 'Ten Commandments for Computer Ethics'. Of the ten, three indirectly relate to safety[6].

As mentioned above, much of the discussion in the literature on computer ethics and in safety-critical literature is still focused on the crucial issues of the co-ordination of diversity and the control of disorder. Hence, we see extensive, and vitally important, discussions on privacy, the right of free [digital] speech in cyberspace, hacking, computer virus control and so forth. What of the hazards posed by a computer system used in the performance of a critically important function? 'Safety' as a moral issue as well as an axiological or value-based issue, has to be discussed more explicitly. Such an approach will help in the interpretation and import of exhortational phrases like, 'at all times take reasonable care', 'take all reasonable steps to keep abreast of developments', 'accept personal responsibility' [IEE 1992] and so on. The consequence of an explicit statement of ethical issues, both moral and axiological, will provide a much needed framework for the specification, design, implementation and ultimate decommissioning of safety-critical systems.


## 4.4 Hare: *ought-to* and being *good*

Richard M Hare, inspired by 'linguistic philosophy', argues for a 'prescriptivist' analysis of moral judgements. He is keen to separate out questions relating to moral judgement from philosophical analysis. He argues that actions are revelatory of moral principles in a way a de-contextualised question about 'what are your moral principles?' cannot reveal. If we accept that it is through our actions we contextualise and elaborate on our moral principles, then Hare's points out that this assumption makes the language of morals a prescriptive one in that 'what are my moral principles?' has to be situated in a specific context and thus becomes reduced or transformed to questions related to action and uttered as 'what shall I do?'

The language of morals has many siblings within a hierarchical classification of 'variants' and 'dialects' of a prescriptive language - the superordinate term. Essentially there are two major types of prescriptive languages, *imperative* and *value-judgemental*. Each of these is further subdivided into two languages. But before we go on to discuss the differences between the two, it is important to be aware of the fact that there are elements of meaning that are stored between imperatives and indicatives.

The simplest form of prescriptive language comprising imperatives is the language comprising *singular* imperatives. Imperatives include military orders; architects specifications, cooking instructions, pieces of advice, requests, entreaties, for Hare all these imperatives can be regarded as commands.

The so-called 'universal imperatives' of ordinary language are not proper universals. [Hare 90:31] Commands are always addressed to someone or to some individual set (not class) of people, hence in a sense one cannot talk about a universal imperative unless it was a moral judgement [ibid:177].

The more complex form of prescriptive language involving imperatives is called the language of universal imperatives or principles - for example, the imperatives contained in the sentence 'never hit a person when he or she is down'.

The prescriptive languages that comprise value-judgemental sentences are divided into *non-moral* and *moral* languages. Hare regards all those sentences containing words like 'ought', 'right' and 'good' express value judgement. Value judgements can be expressed as the grammatical equivalent of statements - 'whatever is expressed by typical indicative sentences', like those used for expressing an opinion that 'you are going to shut the door'. Hare has attempted to relate the logic of 'ought' and 'good', in both non-moral and moral contexts, to the logic of imperatives by constructing a logical model in which artificial concepts, place marking value words of *ordinary language* are defined in terms of a <u>modified</u> imperative mood (see Table 7a)

Table 7a. Hare classification of prescriptive languages

| | | Used for telling someone |
|---|---|---|
| Imperatives (Commands) | Singular<br><br>Universal | that something is the case |
| | | |
| Value-judgements (Statements) | Non-Moral<br><br>Moral | that you think that something is the case |

Value expressions sometimes acquire - by reason of the constancy of the standards by which they are applied - a certain descriptive force. The case of imperatives that are to a high degree 'hypothetical' - like 'if you want to go to the largest grocer in Oxford, go to Tesco's', or 'if you want to break you springs, go on driving as you are at the moment' [Hare 90:35] - helps in reducing imperatives to indicatives. This is because in hypotheticals one says  nothing in the conclusion which is not explicitly said in the premises, except what can be added solely on the strength of definitions of terms. Here an imperative conclusion can be entailed by a set of purely indicative premises.

Hare distinguishes between (and points to the overlap between) the language of statements and prescriptive language on the one hand and between telling a person something and getting the person to believe or do what one has told him or her.

Hare concentrates on trying to avoid the confusion between *good, right* and *ought*, despite the fact that these ethical terms share certain characteristics. Of relevance to our discussion could be the distinction one could make between 'good design', 'right design' and 'obligatory design'. Now, if the professionals involved in a safety argument were to ask an ethical philosopher 'what shall we do?', Hare will offer three different kinds of prescriptives: Types A, B and C together with a 'post eventum' *ought* judgement , and he gives the conditions in which these prescriptives will be appropriate (see Table 7b) :

Table 7b.  Hare's *ought-to* prescriptives

| Type A | Type B | Type C | Type D |
|---|---|---|---|
| Use the starting handle | If the engine fails at once on the self starter, one ought always to use the starting handle | You ought to use the starting handle | You ought to have used the starting handle |
| Get cushions of a different colour | One ought never to put magenta cushions on top of scarlet hostelry | You ought to get cushions of a different colour | You ought to have got cushions of a different colour |
| Pay him back the money | One ought always to pay back money which one has promised to pay back | You ought to pay him back the money | You ought to have paid him back the money |
| SINGULAR IMPERATIVES THAT APPLY DIRECTLY ONLY TO THE OCCASION ON WHICH THEY ARE OFFERED. | THESE PRESCRIPTIONS APPLY TO A *KIND* OF OCCASION, RATHER THAN DIRECTLY TO AN INDIVIDUAL OCCASION | THESE PRESCRIPTIONS APPLY DIRECTLY TO AN INDIVIDUAL OCCASION . | THESE PRESCRIPTIONS ARE *POST EVENTUM* JUDGEMENTS |

Hare has been extensively cited and his work is regarded as a perceptive contribution towards the solution of many fundamental problems of ethics ([Williams 93], [Singer 93]).


## 4.5 Contractarian Models

In these models moral thought is concerned with the kinds of agreements people can make in certain circumstances in which no one is coerced neither any is ignorant.  Such an approach to moral thought is to be contrasted with the insistence in utilitarian systems which focuses on using facts of individual welfare as the basic subject.  The contractarian approach can be traced back to Plato and then onwards to Thomas Hobbes, John Locke, Immanueal Kant and recently to John Rawls.

Contractarian theories hold that 'an action, practice, law or social structure is morally permissible just in case it, or principles to which it conforms, would be (or has been) agreed to by the members of society under certain specified conditions' [Vallentyne 91:3].  Gauthier, its prominent advocate, has argued that a choice made by an agent is rational if and only if relative to his or her belief it is the most effective means for realising the agent's goals.  The philosopher has gone on to argue for his theory under titles, 'No Need for Morality: The Case of the Competitive Market'.  Gauthier and his followers talk about 'constrained maximisation' in an attempt to link 'distributive concerns and moral decisions to a conception of rational interaction that lets individually rational bargainers reach an agreement where everyone foregoes part of his or her potential gain ' [Gaertner and Klemisch-Ahlert 91:163]. Game theoretic techniques to put into operation this novel notion of 'constrained maximisation', in particular, they appear to use 'maximin solutions' for computing a sequence of successive concessions amongst the players.

The derivation of morality from rationality, through game theory, should be of import to the safety-critical community and should complement the diagrammatic techniques and the ethnographic methodology. One extension of Gauthier's theory is 'a more abstract functional approach', the so-called *artificial morality* [Danielson 91]: a complement to artificial intelligence perhaps?

# 5 Conclusions

## 5.1 Safe Design and Repair of Networks

This knowledge of rehabilitating complex networks is distributed literally and metaphorically. In parts this knowledge is personal - usually the privy of experienced design engineers, which is passed on literally by word-of-mouth to the novice designers. Supplementing this experiential, undocumented knowledge is the textual archive comprising text books, learned journals, and manuals, etc. The experiential knowledge interprets and amplifies the textual knowledge. This textual knowledge is the repository of verified and validated experiential knowledge.

The SAFE-DIS project has demonstrated how this word-of-mouth knowledge or experiential knowledge can be archived and used in conjunction with the textual archive. Sometimes the experiential knowledge is used to interpret and amplify contents of the textual archives, and, at other times, the experiential knowledge can be validated and verified using the archives.

This paper has attempted to situate the discussion of safety-critical systems within the broader context of knowledge, language and ethics. In many ways it is not appropriate to talk about conclusions about issues that have attracted the best minds since times immemorial. It is important, nevertheless, that in this age of super-specialisation and the age where some believe that we now have the right technology to replicate the human brain, to look at how knowledge is used by individuals and organisations for building, maintaining and decommissioning safety-critical systems. It is equally important to see how this knowledge is disseminated and to ensure that everyone understands the tension of rapid technological growth and that of public understanding.

The discussion of ethical issues will ensure that hazards posed by a safety-critical system are comprehensible, the risks well understood, and the benefits and costs carefully and openly computed. Our work with the UK water industry, still in a transition from the culture of public corporation to a private sector enterprise, shows that safety issues in the design and repair of urban water-carrying networks are understood as well as is possible. Aspects of this knowledge can be organised within an (intelligent) information system.

## 5.2 Towards a Culture of Safety

The development of safety critical systems is a knowledge-based activity in that there is an extensive use of heuristics, facts, reasoning based on uncertain/incomplete data, rules and meta-rules for building such systems and for building 'knowledge bases' that ensure safe operation of a given safety critical system. This knowledge is, or should be, shared by a range of individuals and organisation. So there is a pool of shared common interests within the safety-critical system communities.

These communities have their own language - a language of safety - that emphasises safe operation and hazard avoidance through the use of an expanding vocabulary that is used in the dissemination of safety knowledge through declarative and imperative sentences.

There is an ethic which governs these communities. The implicit nature of moral principles notwithstanding, there exists within these communities a system based on moral principles that attempts to define obligations, duties and rights, that attempts to provide a framework that may help in distinguishing (good) practices from a totality of practices.

The existence of a knowledge of safety, which is the basis of shared interests within the communities, and that of a set of moral principles, albeit implicitly stated, together with a proposed language of safety used to disseminate and refine the knowledge within a moral framework, constitutes, at least, for us a repository of human action, occasionally mediated through machines, which is socially transmitted. Does this repertoire of socially transmitted action points to the existence of a culture point to the existence of a culture of safety? The evidence we have - learned papers, conferences, ethnographical studies and projects dedicated to safety-critical systems - does suggest that at the very least there is a culture even if it is still evolving.

The constitution of safety knowledge, a clear statement of the moral principles, together with an understanding of the language of safety, will certainly lead to the establishment of a culture of safety complete with its 'kinship structures' and customs for instance.

The 'kinship structures' will help in articulating and establishing (exchange) relationships between, say, the user, the vendors and the operators of a safety critical system rather than the ad-hoc and confused nature of the relationships that exist currently. The 'customs' can be understood at various levels of abstraction: the routine procedures followed by a safety critical community; the rules implicit in the routine; the safety-assuring and hazard avoidance patterns discernible in repetitive acts; the innovations in safety which are either the result of technological development/or the result of novel scientific insights into hazards.

The culture of safety thus will encourage 'safety' as a given rather than safety that has to be imposed.

A safety-critical system developed and established as a result of interactions between the safety kinship structures and through safety customs will indeed be a prescient, presignifying, public good system.

## 5.3 Inclusiveness and Safety-critical Systems

The wider debate about the pros and cons of the private water utilities will be more beneficially held if all parties started by clearly enunciating their moral principles and not to confuse facts with values. If safety related arguments are couched in obscure terminology, by any party in the debate, be it the utilities, consulting engineers, environmentalists, customers, and so on, if these arguments are shrouded in vested interest, then these arguments will convince few and confuse many. This will initially lead to apathy, then to suspicion and then to alienation. We conclude by paraphrasing James Joyce albeit in a context much different to that encountered by his 'young artist', here it can be any one of the parties in the safety argument talking about any or all the other parties:

> 'The language we are speaking is theirs before it is ours. How different are the words *safety, hazard, risk, costs, benefits* on their lips and on ours. We cannot speak or write these words without unrest of spirit. Their language, so familiar and so foreign, will always be for us an acquired speech. We have made or accepted its words. Our voice holds them at bay. Our soul frets in the shadow of their language' (James Joyce 1916 *A Portrait of the Artist as a Young Man.* Herts (UK): Panther Books Ltd, 1964, pp 172. Italicised words are mine - with apologies).

# Acknowledgements

The author is grateful to the SAFE-D*IS* project team at Surrey: Andrew Salway, who co-ordinated the project and worked extensively on the 'language of safety' aspect; Steven Collingham was responsible for programming and for knowledge acquisition; Indrakumaran Selvaratnam investigated the use of multi-agent architectures for safety-critical systems. The SAFE-D*IS Round Table* played a crucial role: Phil Gelder (Severn-Trent Water PLC) was the Chair of the Round Table, and Bob Armstrong (Montgomery Watson Consultants), John Hatley (Walsall Borough Council), Phil High (North-West Water PLC), Jas Mann (Thames Water PLC), and Richard Marshall (Sheffield City Council) were the members of the Round

Table.  Roland Price of Wallingford Software played an important role in the project.  Caroline McInnes was involved in the project administration.  Last but not least, many thanks to  Felix Redmill, who knew better, but waited patiently across many deadlines, thanks to him again.

# References

[Ahmad & Salway 96]   Ahmad, K. & Salway, A. (1996).  'The Terminology of Safety'.  In (Eds.) Klaus-Dirk Schmitz and Christian Galinski.  *Proceedings of 4th International Congress on Terminology and Knowledge Engineering,* Vienna.  Frankfurt: INDEKS-Verlag pp289-297.

[Ahmad 95]      Ahmad, K.  (1995).  'A Knowledge-based Approach to Safety in Design'  In (Eds.) Felix Redmill & Tom Anderson.  *Achievement and Assurance of Safety (Proceedings of the Safety-critical Systems Symposium, Brighton, 1995).*  London: Springer-Verlag Ltd.  pp. 290–301.

[Ahmad & Holmes-Higgin 95]   Ahmad,   Khurshid   and   Holmes-Higgin, Paul.  'System Quirk: A unified approach to Text and Terminology'.  *Proceedings of the Third Term Net Symposium.*  Vienna: Int. Network of Terminology. pp 181-194.

[Aijmer & Altenberg 91]Aijmer, Karin and Altenberg, Bengt.  (1991).  (Eds.) *English Corpus Linguistics: Essays in Honour of Jan Svartvik.*  Harlow (UK): Longman Group Ltd.

[Bell 94]          Bell, R.  (1994).  'IEC Draft International Standard on Functional Safety: Current Position'. *High Integrity Systems.*  Vol. 1 (No. 1) . pp 73-77.

[Bentley *et al* 92a]          Bentley, R., Hughes, J A., Randall D., Rodden T., Sawyer, P., and Sommerville, I.  *Ethnographically-informed Systems for Air Traffic Control.*  (Research Report No. CSCW/3/1992).  Lancaster (UK): Computing Dept., Univ. Lancaster, Lancaster, LA 14 YR.

[Bentley *et al* 92b]          Bentley, R., Hughes, J A., Randall D., and Shapiro, D. (1992).  *Technological Support for Decision Making in a Safety-critical Environment.*  (Research Report No. CSCW/5/1992).  Lancaster (UK): Computing Dept., Univ. Lancaster, Lancaster, LA 14 YR.

[Boose 92]        Boose, John H.  (1992).  'Knowledge Acquisition'.  In (Ed.) Stuart C. Shapiro.  *Encyclopedia of Artificial Intelligence (Vol. 1).*  New York: Wiley-Interscience.  pp 719-742.

[Brazendale & Jeffs 94]            Brazendale, J. and Jeffs, A. R.  (1994).  'Out of Control: Failures Involving Control Systems. *High Integrity Systems.*  Vol. 1 (No. 1) .  pp 67-72.

[Danielson 91]   Danielson, Peter (1991).  'Closing compliance dilemma: How it's rational to be moral in a Lamarckian world'.  In (Ed.) Peter Vallentyne. pp 291-322

[Elliot *et al* 95]   Elliot, John., Brook, Steve., Hughes, Peter., and Kanuritch, Nick.  (1995). 'A Framework for Enhancing the Safety for Advanced Robot Applications'. In (Eds.) Felix Redmill and Tom Anderson. pp. 131-152.

[Fetterman, 93]   Fetterman, David M. (1993) (Ed.) *Speaking the Language of Power: Communication, Collaboration and Advocacy.*  Washington DC (USA) and London: The Falmer Press.

[Fetterman, 93b] Fetterman, David M. (1993) 'Ethnography and Policy: Translating knowledge into action'. In (Ed.) David M. Fetterman.  pp. 156-175.

[Gaertner & Klemisch-Ahlert 1991]      Gaertner, Wulf  and  Klemisch-Ahlert , Marlies. (1991).  'Gauthier's approach to distributive justice and other bargaining solutions'. In (Ed.) Peter Vallentyne . pp 162-179.

[Gerr 42]        Gerr, S.  (1942). 'Language and Science'.  *Philosophy of Science*, Vol. 9.  Pp 147-161.

[Gurr 95]        Gurr, Corin. A., (1995). 'Supporting Formal Reasoning for Safety-critical Systems'. *High Integrity Systems.*  Vol. 1 (No.4). pp 385-396.

[Halliday & Martin 93]  Halliday, Michael A. K. and Martin, John R.  (1993): *Writing Science: Literacy and Discursive Power.* London: Falmer Press.

[Hare 90]        Hare, Richard, M. (1990).  *The Language of Morals.*  Oxford: Clarendon Paperbacks. (Originally published in 1952).

[Hoffman 84]    Hoffman L. (1984): *Seven Roads to LSP.* Fachsprache 1-2/ 1984.

 [Hughes *et al* 95]        Hughes, Gordon., Parey, Deryk., May, John., Hall, Patrick., Zhu, Hong. and Lunn, Dan. (1995)  Nuclear Electric's Contribution to the CONTESSE Testing Framework and its Early Application.  In (Eds.) Felix Redmill and Tom Anderson. pp. 171-187.

[IEE 92] IEE (Institution of Electrical Engineers, UK (1992).  *Safety related systems - Professional Brief.*  London: The Institution of Electrical Engineers.

[Johnson 94]    Johnson, Deborah (1994).  *Computer Ethics (2nd Edition).*  London: Prentice Hall.

[Jones-Lee & Loomes 95]        Jones-Lee, Mike and Loomes, G. (1995). 'Measuring the Benefits of Transport Safety'.  In (Eds.) Felix Redmill and Tom Anderson.  pp. 15-47.

[Kirk 94]        Kirk, Gordon (1994).  'The role of quality assurance in High Integrity Systems'. *High Integrity Systems.* Vol. 1 (No. 1). pp. 79-82.

[Kotonya & Sommerville 94]    Kotonya, Gerald., and Sommerville, Ian. (1994).    Integrating Safety Analysis and Requirements Engineering. (Research Report No. SE/3/1994).  Lancaster (UK): Univ. Lancaster, Lancaster LA 14 YR.

[Mitcham 95]    Mitcham, Carl (1995). 'Computers, Information and Ethics: A Review of Issues and Literature'. *Science and Engineering Ethics.* Vol. 1, pp113-132.

[Moore 73]      Moore, Willis. (1973). 'The Language of Values'. In (Ed.) Ray Lepley. 'The Language of Value'. Westport (CONN., USA): Greenwood Press, Publisher  pp 9-28. (This collection was originally published in 1957 by Columbia Univ. Press, New York)

[Mowshowitz 76]        Mowshowitz, Abbe (1976).  *The Conquest of Will: Information Processing in Human Affairs.*  Reading (Mass., USA): Addison-Wesley Publishing Co.

[Redmill & Anderson 95] Redmill, Felix and Anderson, Tom. (Eds.) (1995) *Achievement and Assurance of Safety: Proc. of the Safety-critical Systems Symposium, Brighton, 1995* London, etc.: Springer-Verlag Ltd.

[Sager, Dungworth & McDonald 81] Sager, J.C., Dungworth, D., and McDonald, P.F. (1981) : *English Special Languages - Principles and practice in science and technology.* Brandstetter Verlag.

[Selvaratnam & Ahmad 95] Selvaratnam, I. & Ahmad, K. (1995). Multi-agent in simulation and modelling. *WOz'95: International Workshop on Oz Programming.* Martigny, Switzerland pp.1-15.

[Seward *et al* 95] Seward, D., Margrave, F., Summerville, I., and Kotonya, G. (1995). Safe Systems for Mobile Robots - the SAFE-SAM Project. In (Eds.) Felix Redmill and Tom Anderson. pp. 153-170.

[Singer 93] Singer, Peter. (1993). *Practical Ethics* (2$^{nd}$ Edition). Cambrideg: Cambridge University Press.

[Thomas 94] Thomas, Muffy (1994). A Proof of Incorrectness using the LP Theorem Prover: The Editing Problem in Therace-25. *High Integrity Systems.* Vol. 1 (No. 1). pp. 35-48.

[Vallentyne ] Vallentyne, P. (1991). *Contractarianism and Rational Choice: Essays on David Gauthier's Morals by Agreement.* Cambridge: Cambridge Univ. Press.

[Walker 95] Walker, Peter M.B. (1995) (Ed.) *Larousse Dictionary of Science and Technology.* Edinburgh and New York; Larousse PLC.

[Weeks and Schensul, 93] Weeks, Margaret R., and Schensul, Jean J. (1995). 'Ethnographic Research on AIDS Risk Behaviour and the Making of Policy'. In (Ed.) David M. Fetterman. pp. 50-69.

[Whitbeck 95] Whitbeck, Caroline. (1995). *Understanding Ethical Problems in Engineering Practice and Research.* New York: Cambridge Univ. Press.

[Williams 93] Williams, Bernard. (1993). *Ethics and the Limits of Philosophy.* London: Fontana Press.

[WRc 86] WRc (1986). *Sewerage Rehabilitation Manual* (2$^{nd}$ Edition). Swindon (UK): Water Research Centre.

[Yen 75] Yen, Ben Chie (1975). *Risk-based design of storm sewers* (Tech. Report no. 141), July 1975. Wallingford (UK): Hydraulics Research Station.

---

[1] The other deliverables of the SAFE-D*IS* include user requirements, specification, design, implementation, and a cost-benefit analysis of the information system.

[2] Selected documents related to SAFE-D*IS* are available through the projects' WWW page at http://www.mcs.surrey.ac.uk /AI/safedis/.

[3] None of these terms were found in the University of Birmingham and Collins Co-Build Publishers 40 million word archive of modern English.

[4] Mev is an abbreviation for million electron volts.

[5] The US-based IEEE covers safety related issues in its core of ethics by asking its members to declare that they will accept responsibility in making engineering decisions consistent with the safety, health, etc., of the public.

[6] The three  of the Washington-based Institute of Ethics 'Ten Commandments' related to safety are *First,* 'Thou shalt not use a computer to harm other people'; *Ninth,* 'Thou shalt think about the social consequences of the program you are writing'; and *Tenth,* 'Thou shalt always use a computer in ways that ensure consideration and respect for your fellow humans' [Mitcham 95 : 121].

# Designing for Safety: Current Activities at the University of Bristol and Future Directions


+G Hughes, +J H R May, *J Noyes
Safety Systems Research Centre, University of Bristol
+Faculty of Engineering, *Department of Psychology


## 1      Introduction

The University of Bristol has recently established a Safety Systems Research Centre (SSRC) to form a focus for safety-related work by bringing together the existing multi-disciplinary strengths from different departments. The Centre has founding sponsorship from the Civil Aviation Authority, Lloyd's Register, Nuclear/Magnox Electric and Railtrack. This funding has been used to develop an underlying research programme on fundamental safety design issues. The programme is intended to provide a point of interaction with other related research covering, aerospace, transport, computer science, communications, medical devices, earthquake engineering, engineering mathematics, process modelling, organisational structure/ management, human factors and psychology. The paper considers what are the important design issues, provides a view of past successes, current activities and future hopes in the area of integrated safety research.

## 2      Safety System Design Issues

In the UK legal and regulatory framework, concepts of safety are articulated as probabilistic concepts of risk. This gives an apparent scientific and quantified image to the systems which contribute to the risk. However, systems continue to change, have increased functionality, become more complicated and in general have to be used before there is any hard empirical evidence of their reliability. There is almost a universal reliance on the design and assessment methodologies in demonstrating that systems are fit for their safety purpose. In recent years there have been some notable 'high-tech' system failures which clearly demonstrate the inability of current methodologies to prevent the 'unexpected'. Such lone data points can have an apparently small effect on mean statistics, but when the root causes are known they are generally strong indicators of a need for design improvement. So are we simply being too ambitious, hiding behind some dubious qualitative estimates of reliability or are we just involved in the gradual process of understanding system design and removing identifiable vulnerabilities? As Engineers we have to believe the latter. So what are the main current identifiable vulnerabilities in system design?

A perhaps simplistic answer is that we do not have or are not prepared to invest in adequate system models. These need to represent a system in adequate detail and be exercised in an appropriate way so that the vulnerabilities are observed prior to operation. Inevitably 'adequate' has to be related to cost, potential benefit, money and time available; qualifying factors which are often summated to define 'practicable'. However, even with infinite resource, there are problems that currently appear to be intractable and it is these which should obviously form the basis of a 'research' programme rather than a 'development' programme. Notable issues are:

- organisation and management of system design and operation and linking this to social factors and pressures;
- understanding human cognitive processes and human performance and linking this into 'balanced' system design;
- providing a formal belief in functional 'correctness' and linking this to probabilistic concepts;
- providing mechanistic failure models and linking these to 'lumped' probabilistic models;
- quantifying test and diagnostic coverage and linking this to probabilistic estimates or measures of performance (reliability/availability);
- reasoning under uncertainty and linking this to improved overall judgements and safety cases.

The first two items recognise the importance of human factors in system design and operation, the latter recognises that there is still great reliance on qualitative 'judgmental' processes. Proof of correctness or testing are the main approaches to demonstrating reliable system design but these are not naturally linked to the required probabilistic estimates. In practice, despite these fundamental limitations to the design process, successful designs are produced. The Bristol contributions are noted in the following review of the above issues.

# 3    Safety Organisation and Management of Risk

It is recognised (see [HSE 93] for example) that organisations achieve safety in system design and operation through the control and avoidance of risks; the co-operation of all those involved; effective communication, and the competence of the individual. The control of risks, or safety management, should be established by a clear allocation of responsibilities for policy formulation, planning and review, the implementation of plans, and reporting of performance. This recognition has led to the social concept of the 'Safety Culture' of an organisation (see for example [IAEA] & [HSE 92]), which is the commitment, maturity and integrity of the organisation and staff with respect to safety issues. The organisation's management should be experienced in dealing with the complexities of safety procedures, safety approval, etc. and have an infrastructure for safety matters that includes:

1. a system for the proper recording and updating of safety related and safety critical equipment and system documentation;
2. a commitment by management to supplying competent, appropriate and knowledgeable personnel with sufficient resources;

3.  a documentation system for safety analysis that records the results of the studies and the management response to the studies.

Whilst these concepts appear to be common sense and supported by hard-learned lessons, there is an inevitable need for a better understanding of the factors at play. This understanding may be essential to sustain an acceptable culture in times of social pressures and change. These are known to produce reorganisation, reductions in resources, loss of experienced staff and knowledge, new production targets, new safety requirements, etc. So far computer models of organisational processes have generally been analysed in terms of a limited scope of activities which have tended to be:
*   pre-defined and contain relatively few decisions which depend on context,
*   fairly linear and sequential,
*   not concurrent or with very few concurrent activities,
*   monitored and managed to ensure objectives are achieved,
*   subject to little change over fairly long periods, and when change is made it is well controlled,
*   time critical,
*   typically high volume,
*   fairly easily represented in terms of some form of work chart, such as a Role Activity Diagram (RAD) or Data Flow Diagram (DFD).

Real organisations are dynamic, highly concurrent and not amenable to simple description and pre-definition. For example the definition of a formal procedure does not guarantee that the process is performed in this way in practise. New models are needed to identify and study the important factors influencing the performance of organisations.

Work on the development of new organisational process models has been performed by the Systems Research Group, within the department of Civil Engineering, based on the concept of "Interacting Objects" [Blockley 95]. The concept has a potentially wide application and it has, for example, been used successfully to model physical processes. In a process model, with human involvement, message input is filtered by perception, reflected upon and then acted upon.   This, in turn triggers perception-reflection-action sequences in other interacting object (IO) process roles. This concept called the Reflective Practice Loop (RPL) was introduced [Blockley 92] to model the basic processes in problem solving and decision making in science and engineering.  The RPL is being applied to the modelling of roles within an organisation.  Roles receive messages from other roles, the message is filtered through a perception module, passed into a reflection module which operates on it and outputs a message through an action module to another role or to the outside world.

The RPL is being made self replicating by enabling it to create sub roles, identical in structure but void of content until occupied by Man or machine and to which it can then delegate some of its responsibilities.  This initiates a further process in which the two roles will negotiate new objectives. Progress towards these objectives can be monitored by both roles. It follows that in their turn sub roles can create further sub roles generating a hierarchy of responsibilities. The result is a series of delegation and feedback RPLs. The RPL is being programmed into a simple role object. The minimum requirement of the simple role object is that it should be capable of

- interpreting messages,
- reflecting upon that information so that it can decide on action,
- sending messages to neighbours,
- replicating itself,
- accessing and using tools such as spreadsheets and databases.

The internal structure of the Role Object directly corresponds to the RPL. The reflection aspect is course difficult as it is a function of human cognition, experience, knowledge, support systems, rules, constraints, targets etc. The role object is also able to create and delegate responsibilities to sub-roles, monitor the progress of sub-roles and send monitoring information to the role to which it is accountable.

The model of the structure of the organisation is developed by 'growing' the self replicating unit through all of the roles within that organisation. It is important to note straightaway that the resulting hierarchy is not a hierarchy of authority (i.e. the structure through which legitimate power is exercised), except for that type of authority which is associated with higher levels of responsibility [Handy 85]. The hierarchy should not be seen as the setting up of an authoritarian power structure. The culture of the organisation is a separate issue largely set by senior management. At one extreme it may be hierarchical (as in the military) or loose and flat in the sense that all workers are colleagues (as in Universities). The hierarchy suggested here is one of delegated responsibilities which can be, but is often not in practice, set up independently of the organisational culture.

Of course the objectives set under any given role may not be met for various reasons. For example it is possible that the objectives were inappropriate. In such a case the delegating role needs to initiate a re-negotiation of objectives in order to improve performance. The system is therefore one of a double loop learning where the delegating role not only monitors the performance of the sub-role but also evaluates its effectiveness and helps to diagnose remedial actions. This view of management and leadership is in accord with [Senge 90] who argues that the new view of leadership is that leaders are designers, stewards and teachers. They are responsible for building organisations where people continually expand their capabilities, clarify vision and improve team learning i.e. they are responsible for organisational learning.

The motivation for this work is that appropriate support for business processes within a company will increase quality, enable business re-engineering and hence reduce the risk of failures and accidents.

# 4    Human Cognition and Performance

The concept of modelling an organisation is only feasible if it is possible to represent the 'reflection' process of the human cognitive performance observed in the context of the constituent simple or restricted roles.  A similar problem exists for all contexts found in the design and operation of human-machine systems.  One of the overall objectives for the design of such systems is the achievement of an acceptable 'balance' between automation and human action; thus, ensuring that the human is (and feels) in overall control.  Research has shown that the degree of control over the situation perceived by the individual is an important determinant of risk perception [Mearns and Flin 96].

Although automation is generally welcomed in terms of enhancing system operation and increasing reliability, increasing the degree of automation is not guaranteed to improve system operation and reliability.  Firstly, it transfers increased dependence to automatic system design and maintenance (human activities with associated error potential), and secondly, it tends to reduce the operator's understanding of the system as well as possibly reducing his/her involvement to that of a monitoring role, perhaps even to the point of boredom.  It is known that humans are poor at vigilance tasks, especially when checking for low frequency events [Wickens, 84].

The automation issue has been investigated during some recent work carried out at Bristol in conjunction with British Airways and Smiths Industries Aerospace on the application of model-based reasoning techniques in the development of an advanced warning system for civil aircraft (see, [Noyes *et al.* 95]).  Survey work with flight deck crew indicated that greater automation and improved system reliability generally result in a reduction in the extent of the crews' interactions with the aircraft systems.  This reduction in 'hands on' operation tends to lead to a decrease in the crew's knowledge and experience of detailed system function. Although all the relevant flight information is present within the system, there is no procedural or operational need to interact with it, and this lack of interaction results in crews having less need to cross-check and discuss aspects of the flight with each other.  This can result in a subsequent loss of situation awareness - a concept described as 'peripheralisation' by [Satchell 93] and others.  In the aviation world, this aspect of automation is recognised as being of increasing concern [James *et al.* 91].  However, situation awareness is not simply awareness of system states, but also extends to include the interpretation of data pertaining to these systems [Pew, 94].  For example, some current civil aircraft warning systems are programmed to attract the crew's attention only when parameters pass out of limits, i.e. beyond pre-determined fixed thresholds.  Consequently, no failures mean that there are no distracters for the crew, but also no information.  It has been pointed out by [Wiener 87] that human operators must continually exercise their capacity to analyse, seek novel solutions and extrapolate beyond the current situation, and automation as a general rule does not always allow or encourage this.

A further issue concerns the extent to which automated systems should override the decisions of the human operator. A number of aircraft incidents have highlighted this problem, e.g. the accident report for the A320 which overrun the runway in Warsaw in September 93 indicated that the overrun would probably not have happened if the aircraft's protection systems had "not delayed deployment of the lift dumpers and reverse thrust, even though the pilot selected them" [Learmount, 95, p. 24].

In summary, an extreme view of automation is that it will eventually result in systems that no one will understand [Perrow, 84]. Furthermore, in the event of a major malfunction, flight deck crew could be required to transfer very rapidly from simply 'supervising' a wholly automated operation to manually handling a severely disabled aircraft. The degree of automation is therefore an important (as well as being a controversial) consideration in the next generation of flight deck warning systems [Daly *et al.* 94].


# 5   Formal Correctness and Probability

The ADEPT project is a research project in the Department of Computer Science at Bristol University, intended to run for 3 years initially, which aims to:

1. develop industrial-strength implementations of declarative languages and their associated program development tools;
2. apply these advanced languages and tools in suitable application areas to confirm their effectiveness compared with existing technology; and
3. help facilitate the much wider use of these languages and tools in industry and commerce.

The vast majority of industrial and commercial applications make no use whatever of declarative programming technology. The primary aim of the ADEPT project at Bristol is to address this deficiency and contribute to the industrialisation of advanced software production techniques which are currently being developed at Bristol University and other university and industrial research centres around the world. The core of the project is the declarative, general-purpose programming language Escher [Lloyd 95] which has been under development at Bristol for nearly two years. Escher will provide a vehicle for research and experimentation, and also provide a common technical basis for all researchers in the project.

Building on the considerable experience and skills in the Computer Science department at Bristol, the project is centred around a plan to develop an industrial-strength implementation of Escher, complete with appropriate programming development tools such as program optimises, debuggers, and compiler-generators. Suitable industrial applications will then be selected and, in collaboration with industrial partners, solutions to these applications will be implemented and evaluated using the tools developed. To supplement the value of this work for safety applications, a parallel probabilistic safety case will be produced for use with the new language/tools as a means of focusing attention on the weaker aspects of the development.

Further information about the ADEPT project is available at http://www.cs.bris.ac.uk/~jwl/.

# 6 Failure Modelling

Conventional failure models come in many forms (e.g. failure mode and effects analysis, fault trees, event trees) but all rely greatly on human understanding and knowledge and provide rather informal ways of abstracting failure models from a system. Safety related systems usually have unique qualities, they are seldom off-the-shelf, and therefore safety justification cannot rely on previous experience with the system as a whole. Ideally then, a model should provide failure analysis prior to use of a system in the real world. Failure models must describe the failure behaviour in terms of attributes of a system which are measurable at that time. Perhaps the best known example is standard fault tree analysis in which system failure is described in terms of the failure behaviour of its components, which is understood on an individual basis and has been measured.

In all failure models there are imposed limitations or "discrimination" in terms of event types or failure states considered in order to make solution tractable. For example, fault tree analyses usually adopt Boolean logic involving only two component states (normal or failed) with causal links restricted to AND/OR gates. These clearly fulfil the strong human desire to understand empirical phenomena in terms of cause-effect relationships and provide simple structures allowing relatively efficient computation. The structural and logical simplicity is often earned at the expense of uncertainty and judgement in identifying the vast variety of possible failure events, discriminating 'significant' events, finding the causal probability data that is needed to support the model and the assumption of independence of root causes.

One reason for performing failure modelling is the greater understanding of a system that results. The identification of critical failure modes, or of critical areas of software code, so that they can be 'designed out' or otherwise given special attention, remains an important role for such models. However, the UK legal safety framework is based on risk. Thus an additional crucial aim is the quantification of failure likelihood for these systems which is a severe complicating factor for the models.

It may be that failure models are diverse because failure processes are not yet fully understood. The current state of the art certainly includes diverse approaches. To explain the areas in which SSRC research contributes, a classification of failure models is helpful. In the following, a failure is an outcome caused by a component or system which is contrary to requirements, and a fault is the physical condition of the component or system which causes failure. A fundamental distinction concerns the type of failure being modelled. Two types of failure are often distinguished: systematic failure and random failure. Systematic failure of a system is caused by faults present prior to any degradation in that system. In this sense they are built-in, although this need not imply intentionally. Whilst both types of failure are the subject of current research, models for systematic failure are a particular problem. In fact there are no universally agreed solutions. Systematic failure studies tend to focus on software and on computer processor electronics, simply because the very high complexity of those systems increases the likelihood of design error. Most recent work by SSRC staff has been on systematic failure modelling. In contrast, random faults are not present in a new system/component in the sense that the system/component will initially behave as required in a given task. Instead they involve a process of alteration (degradation) of system components, and the processes of this change are incidental to the specific required system function. By incidental it is meant that whilst these processes are inevitable, they are not built-in in a positive sense - as part of the attempt to make the system achieve its required function - but rather they are a side-effect. Random failure is the traditional subject of reliability analysis, for which some good mathematical models exist.

A second distinction can be drawn between correlative and logical models. This distinction is orthogonal to the random - systematic distinction. In an correlative model, the presence of system failures or faults is correlated with a number of influencing variables. Observation of the actual values of the influencing variables for a system, results in a particular likelihood distribution for, for example, the number of faults present. In a logical model, system failure is not defined in this uncertain manner, but rather logically in terms of other events. For example, a fault tree describes system failure in terms of component failure; if a failure of all components in a cutset occurs, system failure occurs with certainty.

An example of a random, correlative models are the statistical models for hardware component failure rates. Fault trees, event trees, block diagrams etc. can all implement random, logical models. Systematic, correlative models include: the FASGEP inner model (see section 6.3); the COQUAMO model which relates certain software metrics to software quality [Kitchenham & Walker, 89]; and the single urn model of software failure probability on demand estimation [Miller et al, 92]. Systematic, logical models, although advocated by some [Leveson et al, 91], are rare. Complications arise when this approach is attempted , particularly if a quantitative approach is required.   However, new hybrid logical/correlative quantitative models for systematic failure being researched at the SSRC include logical techniques (section 7.0), and the FASGEP outer model is another example of this kind (section 6.3).

## 6.1 Decomposition-based failure models

Logical models are based on some decomposition of a system into a set of components. Current correlative models often treat the system en-bloc, as is the case with the single urn (systematic failure) model of statistical software testing [Miller et al, 92], and the COQUAMO model. For the purposes of discussion, a component will be taken to mean any constitutive part of a system, including human activity. Failure modelling with decomposition is attractive because it seems reasonable to assume that it is easier to understand and gather data about component behaviour than the behaviour of a system as a whole. Furthermore, modelling a system as an interacting set of components is intuitive and familiar. However, research into decomposition-based failure modelling has always faced serious difficulties which arise because it differs in significant ways from the modelling of correct system behaviour. In a correctly working system the intended functions of components are usually quite specific and restricted. Furthermore a single component interacts with a relatively small number of neighbour components in an understandable fashion, and this is an artefact of the way systems are built - it being difficult to conceive systems which work using complex component interactions. It is these simple component interactions which are traditionally used to construct a failure model, based on formalisms such as fault and event trees [McCormick, 81] [IEC 1025] and block diagrams [Beasley, 91] [IEC 1078], describing the dependence of the success (failure) of system function on the success (failure) of component function. The way in which decomposition is used depends on whether component failure modes are known. Its use is straightforward provided component failure modes are known, as explained in section 6.2. A different and more subtle use of decomposition is needed when failure modes are unknown, as is the case for systematic failure (section 6.3).

Experience has shown that unanticipated routes to system failure are not uncommon [McDermid, 93]. The interesting and problematic property of these unexpected failures is that they often concern collections of components failing together in a non-random fashion. The study of component subsets, as opposed to "autonomous" components, is inherently more complex because the number of subsets is vastly greater then the number of components. Discrimination of the important subsets is key but is an extremely taxing problem in general.

## 6.2 Failure modelling with decomposition and known failure modes

Traditional formalisms only address a part of the failure modelling problem. Specifically, given a component cut set for system failure, there remains the difficult problem of anticipating routes to failure of this cutset. Overlapping random failure of components in a failure cutset is the traditionally studied route to system failure. Traditional methods of constructing failure models adequately identify such routes because there is no need to model relationships between component failures.

Unfortunately, inter-related component failures cannot be ignored. This problem has long been recognised; traditional reliability analysis knows these non-random phenomena as common cause failures [McCormick, 81]. However, a distinction must be made. A failure of an electricity supply might be viewed as a type of common cause failure since it causes a collection of components to fail. This is not difficult to anticipate and model within the traditional approaches because failure is due to a failed supply over a familiar component interaction path of the correctly working system. The real modelling difficulties are due to common cause failures which involve unintended component interaction paths. Cascade failure provides one example. It is not simply the mathematics and reliability data requirements of cascade failure which present difficulties, but more fundamentally the anticipation of cascade event trains. The reason for this is that random component failure is precisely a change in component function, and therefore can introduce entirely new, unintended interactions between components e.g. loss of fluid from a burst pipe causing electrical failures, or misinterpretation of alarm information by human plant operators inducing inappropriate interventions. That is, failure does not only alter component interactions over existing interaction paths, but can create new interaction paths - including interaction between components which did not interact prior to failure. Furthermore, it is not necessary for a component to fail in order to cause unintended interactions. A normal working property of a component, which may or may not play an intentional role in system function, can cause failure of other components e.g. natural vibration in a pump causing fatigue in nearby structures. The identification of unintended component interactions is therefore a key task in safety analysis, but methods and tools to provide significant help with this task are lacking. The type of modelling required would mix top-down (e.g. fault trees) and bottom-up (e.g. failure modes and effects analysis [IEC 812]) techniques, focusing on unusual or unintended interactions and the problem of reducing the complexity of the problem of considering all possible interactions. The SSRC will seek to identify such methods, and build tools to implement them. One candidate technology would be case based reasoning techniques. Consideration will also be given to an additional and different modelling problem which arises when failures can be caused by degraded performance in a number of components, none of which on their own would cause system failure.

## 6.3 Failure modelling with decomposition where failure modes are unknown (systematic failure)

In some ways systematic failures present a more difficult problem then traditional failure modelling. Historical information on component failure modes is not normally available, so that traditional failure models cannot be used in the same way. However, interestingly the modelling difficulties stem from the same reason, namely, the necessity of considering related failures of collections of components and not just failure of components in isolation. The ideas of incorrect interactions along intended paths and incorrect interaction paths (an example of the latter being side-effects in software) still apply. In contrast to the random case, incorrect interactions along intended paths are no easier to model than interactions along unintended paths. This is because they are not necessarily due to unintended performance of a component as is the case with random failure. Components may be working to their requirements and interacting entirely as intended but still produce an incorrect result; the problem being incorrect conception by the system designer. For systematic failure then, it is certainly not possible to model system failure solely in terms of independent failure of individual components. On the other hand there is an advantage over the case of traditional failure modelling, namely, better control over the relevant aspects of the system environment - those which affect systematic behaviour. As a result, system testing becomes a powerful tool. It is possible to test component collections directly, including the whole system, and it is not necessary to rely on testing individual components as we have to in the case of random failure. The result is that systematic failure modelling has a different emphasis from traditional approaches. Models are not concerned with analysis of particular types of failure, but rather with demonstrating that faults or failures of any type are rare.

Different approaches are possible. A common categorisation is according to the source of evidence used in the model. Evidence of system integrity can be collected either from testing of the product (e.g. statistical software testing) or testing of the process which created the product. SSRC is researching both approaches, based on established work of SSRC staff. Firstly, statistical software testing is being used as a vehicle to investigate testing of systems (products) to guard against systematic failure. New statistical models have been derived. The aim is to formalise test requirements based on system risk targets. This work is described in section 7.0. Secondly, the process-testing approach used in the FASGEP project is to be developed at the SSRC and is briefly described below.

The DTI/EPSRC FASGEP project developed an approach based on the idea that the systematic faults embedded in software are introduced during the development process by human activity. It is assumed that the likelihood of fault introduction depends on features of the development process such as its size/complexity, and factors such as the experience and quality of development staff and the difficulties faced by them during development. The output of the model is a probability distribution describing the predicted number of faults introduced up to any point in a development process. It is based on graphical probability models [Pearl, 88] [Lauritzen & Spiegelhalter, 88]. Currently, the purpose of the model is to guide a software development process prior to testing. In particular, the model can be used to decide when reviews become necessary, based on the predicted fault intensities. However, a further objective of the work at the SSRC is to formalise the evidence associated from different software development and assessment methods which supports a safety case, linking this to empirical failure/reliability evidence. For example, FASGEP model output might be used to condition a new statistical software failure probability estimator, thus combining evidence from both process and product testing.

The model has two major components.
i.    A logical "outer model" which decomposes the development process into a directed acyclic graph of "atomic processes." The outer model aggregates the faults introduced in all atomic processes, taking into account the numbers of faults

identified in reviews and the possibility that corrective rework also has the potential to introduce faults.

ii.    A correlative "inner model" which correlates measured (i.e. tested) 'attributes' of an atomic process with its potential to introduce (for design processes) or identify (for review processes) faults.

The DTI/EPSRC FASGEP model is described in detail elsewhere [May et al, 93] [Cottam et al, 95]. A final model for public distribution and use has recently been commissioned by the DTI and completed by Nuclear Electric with SSRC support. There are real methods for calibrating the model using attribute measurements and fault data from real projects [Spiegelhalter & Lauritzen, 90]. However, the data requirements for the model in its current form are too large, despite the potential for almost unlimited numbers of projects. It is not clear that smaller models would lack accuracy, so a future research direction will investigate the consequences of reduced nets with more practical data requirements.

# 7    Testing and Reliability Estimation

As reasoned in section 6.3, system testing has the potential to be a key defence against systematic failure. For safety related systems, any faults identified by testing must be corrected so that the testing problem of greatest interest is inference of reliability from a period of failure free operation. For software this form of testing is sometimes called statistical software testing (SST). The SSRC is using SST as a case study for the general problem of system testing against systematic faults.

## 7.1 Statistical inference from testing

Current accepted SST [Littlewood & Wright, 95] [Miller et al, 92] [Thayer et al, 78] are not logic models as defined in this paper. It is interesting to understand why this is the case. The main reasons are that the greater control over the system environment (discussed in section 6.3) makes en bloc testing of a system possible, and also that researchers have not been able to discover the significance of system decomposition for SST. This does not mean that it has no significance. Intuitively, ignoring software structure is throwing information away. Consider two programs: one very short and simple, the other hugely long and complex. Given we have this information, is it reasonable to assume that, say, N tests produces the same reliability estimate in both cases? Should the estimate really be immune to the diversity of function and/or patterns of code execution used to achieve that function?

There has been some work on logical models of software testing, an early example was provided by Littlewood [Littlewood, 81]. It could also be argued that the multiple urn model of Miller et al is partly logical in nature. [Miller et al, 92]. The Littlewood work decomposed the code itself, whilst Miller et al worked on decomposition of the software's input space. New models have been developed by SSRC members which use both of these types of decomposition [May & Lunn, 95a]

[May & Lunn, 95b]. These models are a first attempt to investigate the two questions above. Another interesting perspective on these models is that they combine notions from the previously disjoint fields of statistical testing and test adequacy measurement (e.g. code coverage measures).

## 7.2 Environment simulation

Environment simulation is another important area of testing research. Meaningful system testing must model the system environment realistically, and testing should "cover" the range of possible environment behaviours in some sense. The SSRC can draw on experience of its staff on the CONTESSE DTI/EPSRC project [May, Hughes & Lunn, 95], but also on recent research by Bristol University's Civil Engineering Systems Group on Interacting Objects Modelling (IOM) [Blockley, 95]. IOM appears particularly attractive for modelling safety related system environments for several reasons. Firstly, such environments are typically man-made. They work by connecting components of known functionality together in prescribed ways. IOM allows simulations to be constructed in an analogous way to the real environments. Secondly, component simulations may be reused. Thirdly, IOM will facilitate convenient simulation of environment failure modes. This last point is important because safety related systems are commonly required to perform when their environments enter failure modes. With an IOM approach, simulators of environment failure modes may be efficiently constructed by substituting a simulator of a correctly working component with a simulator of that component in a failure mode.

For SST, it is necessary to build environment simulations which replicate the probabilistic properties of the environment i.e. the software's operational input distribution [Musa, 93]. There is little literature on this subject, and methods seem to depend on the software application being tested. SSRC staff have worked on operational distribution construction in the context of the Sizewell B Primary Protection System [May, Hughes & Lunn, 95]. Future work on this subject will investigate the effects of uncertainty in an operational distribution on final reliability estimates i.e. a study of the sensitivity of these estimates to changes in the operational distribution.

## 7.3 On-line Diagnostics

The fault coverage afforded by an on-line diagnostic technique is a concept of fundamental importance in the design of safety systems which have identifiable fail-safe-states. The ability to reveal a large fraction of all potentially dangerous faults and put the system into a safe state can be used to ensure reliable performance of a safety function. The concept is detrimental to system and plant availability but in the safety context, this is of secondary importance. The concepts of fault tolerant design based on the use of switchable redundant elements and recovery blocks can be used to maintain plant availability. Quantification of the Coverage Factor provided by diagnostics is thus a potential method of quantifying safety system reliability.

At present there is no overall methodology which combines analytic assessment methods such as Functional Block Analysis with the design of on- line diagnostic techniques. The aim of such a methodology should be to quantify a design coverage factor which can be the basis of an availability/reliability claim rather than the token gesture that is common. The importance of the integrity of the diagnostic method is also apparent and, logically, designs should grow from an element which is totally self-checking.

SSRC will explore the design of fault tolerant and fail-safe systems to incorporate elements which are totally self checking and could form a safety kernel together with the quantification of coverage factors for partially self-checking elements. The approach could use a combination of analytic methods and coded information techniques. There is an obvious synergy with the work proposed above on structure based statistical models.


# 8     Current Safety Applications

The generic topics related to safety system design, outlined above, support a number of important application areas at Bristol.

## 8.1 Medical Devices

In the field of surgery new technology combining software, actuation and sensors offers great benefit in the form of invasive tools. Such devices are safety critical systems and so far have only been demonstrated in research circles on patients in the operating theatre. This has been a significant step forward. These devices have used basic software, simple sensing devices and simple actuation techniques to carry out procedures that have parallels to an NC machine where the position and trajectory is well defined from previous data. Their movements can be verified and all decisions made by the surgeon prior to the operation. It has been important to select suitable procedures to demonstrate these new tools and so far procedures have been where tissues are considered stiff such that they do not deflect significantly under tool action. In these circumstances scan data of the target position can be relied upon to plan trajectories.

The benefits of machine action over manual tool control are accurate motion and the ability to remain poised in a fixed position. In addition it is possible to make a machine react more rapidly to changes in sensory data and to move with much more steadily micro tool actions. While there are benefits there are also challenges. Safety is paramount and therefore the level of risk has to be justified and well validated. As the field is new and growing there is an urgent need to guide researchers and producers of new tools through an acceptable development procedure for these devices. This will enhance the confidence of manufacturers and researchers to progress with new ideas and products that can be used safely to benefit many patients.

In the future there is an important aspect that has to be recognised. To obtain the full benefits of this technology in surgery, decision making by machines is necessary. Future tools will make decisions and move accordingly to sensory data rather than along trajectories with velocities that are pre planned and there are some examples already in research. One such device rapidly approaching clinical trials is the stapedotomy micro-drill at the University of Bristol. The device drills through a very thin flexible bone structure that deflects significantly under tool action and the aim is to minimise protrusion of the drill bit beyond the far surface. Tool movement is small and there is an interest in precision. The drill is 0.6 mm in diameter and drills through the bone which may vary in thickness from 0.2mm to 2.5mm. the compliance of the bone and the drill cutting condition vary between each case. the drilling machine developed is computer controlled and measures force, torque and the drill feed relative to the bone. From this it is able to interpret the state of the drilling process and to determine the position of the unknown far medial surface of the bone prior to breakthrough of the drill bit. It is then able to decide how to move the drill bit to minimise protrusion of the tip beyond the far surface. This ensures a higher level of safety than is possible manually when drilling or when using other manual tools.[Brett et al 95].

## 8.2 Advanced Transport and Avionics

Separate papers are presented at this symposium on the safety issues associated with the design of new Personal Rapid Transport Systems [Lowson and Medus 97] and the use of functionally dissimilar monitoring to increase software integrity [Johnson 97]. In addition, a study on Reconfigurable Integrated Modular Avionics is being sponsored by the Civil Aviation Authority to identify suitable architectures, and define the reconfiguration mechanism.

## 8.3 Earthquake Engineering

The Earthquake Engineering Research Centre at Bristol co-ordinates the EU's 4m ecu programme of earthquake research and facilities include the EPSRC Earthquake Simulator (or shaking table). Current research includes the seismic and dynamic analysis of dams and fluid retaining structures, long span bridges, buildings, soils and foundations and safety and vulnerability analysis. Its commercial arm provides seismic qualification of computer/electronic and electro-mechanical systems.[Severn 95].

## 8.4 Software diversity

Finally, a new project has been set up to use failure modelling and testing to investigate a particular technique for software safety, namely software diversity. Methods for achieving diversity will be derived and tested for their effectiveness. For example, one experiment will test to see if two diverse systems with a pfd (probability of failure on demand) of $10^{-2}$ can be used to in parallel to create a $10^{-4}$ pfd system. The work will be based on four diversely developed versions of realistic complexity boiler control software.

## Acknowledgements

The authors wish to thank Professor D. I. Blockley, Professor R. T. Severn, Professor J. W. Lloyd, and Dr P. N. Brett for their contributions to the paper.

## References

[Beasley 91] Beasley M. Reliability for Engineers, Macmillan, London 1991

[Blockley 92], Blockley D.I. Engineering from Reflective Practice, Research in Engineering Design, 4, 13-22 1992

[Blockley 95], Blockley D. I. Computers in Engineering Risk and Hazard Management, Archives of Computational Methods in Engineering, Vol. 2,2,67-94, 1995.

[Brett et al 95], Brett P.N., Baker D.A. and Blanshard J.A. Precision control of an automatic tool for micro-drilling a stapedotomy, ProcIMechE, part H, vol 209, pp255-262, Dec 1995.

[Cottam et al 94] Cottam M., May J. et al Fault Analysis of the Software Generation Process - The FASGEP Project, Proceedings of the Safety and Reliability Society Symposium: Risk Management and Critical Protective Systems, Altrincham, UK October 1994

[Daly et al 94] Daly K., Jeziorski A. and Sedbon G. Intelligent conversation, Flight International, 24-30th August, 25-27 1994.

[Handy 85] Handy C.B. Understanding Organisations, 3rd Ed, Penguin Books, London, 1985.

[HSE 92] Organisational Management and Human factors in Quantified risk Assessment, Reports 33/1992 & 34/1992.

[HSE 93] Successful Health and Safety Management

[IAEA] The International Atomic Energy Agency Guide to Safety Culture, (Safety Series no. 75-INSAG-4; STI/PUB/882.)

[IEC 812] Guide to FMEA and FEMCA/ BS 5760 Pt 5. 1985.

[IEC 1025] Guide to fault Tree Analysis/ BS 5760 pt 7, 1990.

[IEC 1078] Guide to the Block Diagram Technique/ BS 5760 Pt 9, 1991.

[James et al 91] James M., Mcclumpha A., Green R. Wilson P. and Belyavin A. Pilot attitudes to automation, In Proceedings of the Sixth International Symposium on Aviation Psychology, (Ohio State University, Columbus), 192-197, 1991

[Johnson 97] Johnson D.M. Increasing software integrity using functionally dissimilar monitoring, This Volume, 1997.

[Kitchenham & Walker 90] Kitchenham B.A. and Walker J.G. A quantitative approach to monitoring software development, Software Engineering Journal, Jan 1989.

[Lauritzen & Spiegelhalter 88] Lauritzen S.L. and Spiegelhalter D.J. Local Computations with Probabilities on Graphical Structures and Their Application to Expert Systems, J. Royal Statistical Society B, v50 n2 1988

[Learmount 95] Learmount D. Lessons from the cockpit, Flight International, 11-17th January, 24-27, 1995.

[Leveson et al 91] Leveson N.G. Cha S. and Shimeall T.J. Safety verification of Ada programs using software fault trees, IEEE Software SE-17, July 1991

[Littlewood 81] Littlewood B. Software reliability model for modular program structure, IEEE Trans. on Reliability v R-30 1981

[Littlewood & Wright 95] Littlewood B. and Wright D. Some conservative stopping rules for the operational testing of safety-critical software, IEEE Trans on Fault Tolerant Computing Syposium, pp 444-451, Pasedena, 1995.

[Lloyd 95] Lloyd J.W. Declarative Programming in Escher, CSTR-95-013, Department of Computer Science, University of Bristol, 1995.

[Lowson and Medus 97] Lowson M.V. and Medus C. An initial study of Personal Rapid Transport (PRT) Safety, This Volume, 1997.

[May et al 93] May J. et al. Fault Prediction for Software Development Processes, Proceedings of Institute of Mathematics and its Applications Conference on the Mathematics of Dependable Systems, Royal Holloway, Univ. of London, Egham, Surrey 1-3 Sept. 1993

[May, Hughes & Lunn 95] May J., Hughes G and Lunn A.D. Reliability Estimation from Appropriate Testing of Plant Protection Software, IEE Software Engineering Journal, Nov. 1995

[May & Lunn, 95a] May J.H.R and Lunn A.D. New Statistics for Demand-Based Software Testing, Information Processing Letters 53, 1995

[May & Lunn, 95b] May J.H.R & Lunn A.D A Model of Code Sharing for Estimating Software Failure on Demand Probabilities, IEEE Trans. on Software Engineering SE-21(9) 1995

[McCormick 81] McCormick N.J. Reliability and Risk Analysis, Academic Press, New York 1981

[McDermid 93] McDermid J. Issues in the development of safety-critical systems, in Safety-critical Systems: current issues, techniques and standards, Eds. F Redmill & T Anderson, Chapman & Hall, London 1993

[Mearns and Flin 96] Mearns K. And Flin R., Risk perception in hazardous industries, The Psychologist, 9(9), 401-404, 1996

[Miller et al 92] Miller W.M. , Morell L.J., Noonan R.E., Park S.K., Nicol D.M., Murrill B.W. and Voas J.M. Estimating the probability of failure when testing reveals no failures, IEEE Trans. on Software Engineering v18 n1 1992

[Musa 93] Musa J.D. Operational profiles in software reliability engineering, IEEE Software 10(2) 1993

[Noyes et al 95] Noyes J.M., Starr A.F., Frankish C.R. and Rankin J.A. Aircraft warning systems: Application of model-based reasoning techniques, Ergonomics, 38(11), 2432-2445, 1995

[Pearl 88] Pearl J. Probabilistic Reasoning in Intelligent Systems, Morgan Kaufmann, San Mateo 1988

[Perrow 84] Perrow C., Normal Accidents: Living with High Risk Technology, (Basic Books, New York), 1984.

[Pew 94] Pew R.W. Situation awareness: The buzzword of the '90s, CSERIAC Gateway, 5(1), 1-16, 1994

[Satchell 93] Satchell P. Cockpit Monitoring and Alerting Systems, (Ashgate, Aldershot), 1993

[Senge P, 90] Senge P. The Fifth Discipline: The Art and Practice of the Learning Organisation, Century Business Books, 1990

[Severn 95] Severn R.T. The European Shaking Table Programme, Keynote Address, SECED Conference on European Design Practice, Chester UK, September 95, Elesvier.

[Spiegelhalter & Lauritzen 90] Spiegelhalter D.J, and Lauritzen S.L. Sequential updating of conditional probabilities on directed graphical structures, Networks 20, 1990

[Thayer et al 78] Thayer R., Lipow M and Nelson E. Software Reliability, North-Holland, Amsterdam 1978

[Wickens 1984] Wickens C.D. Engineering psychology and human performance. Columbus, Ohio: Charles E. Merrill, 1984.

[Wiener, E.L. 1987] Wickens E. L. Management of human error by design, In Proceedings of the 1st Conference on Human Error Avoidance Techniques, Paper 872505, (SAE International. Warrendale, PA), 7-11, 1987.

# Using a Layered Functional Model to Determine Safety Requirements


J U M Smith
Nairana Software Ltd.
London, United Kingdom


## Abstract


This paper describes the use of a layered functional model to obtain safety requirements for a data processing system supporting air traffic control. The model was used as the basis for a FMECA analysis to identify system hazards. The hazards were then turned into probabilistic safety targets for the system.


## 1. Introduction

This paper discusses the problems involved in determining the safety requirements of a large and complex data processing system used to support air traffic control.

The system was the National Airspace System (NAS), one of the principal data processing systems used by the National Air Traffic Control Services Ltd to support the air traffic control service over England and Wales. Although NAS was an existing system with a history of successful and safe service, it is was being modified to provide new functionality and to support new interfaces. It was therefore felt that a full analysis should be carried out of the whole system so that the effect of the proposed changes could be judged. This involved first creating safety requirements for the existing system, retrospectively, to provide a baseline against which the proposed changes could be judged.

Historically much attention has been focussed on the availability of NAS. NAS is required to be operational and available around the clock, 365 days of the year. Comprehensive reports are produced daily and monthly on the percentage of down time being experienced, see table 1. However as Nancy Leveson has pointed out safety is a different and distinct system property [Leveson 95]. In the case of NAS safety turns out to be more closely related to data integrity than system availability.

| Year | 1990 | 1991 | 1992 | 1993 |
|---|---|---|---|---|
| Availability | 99.97 | 99.98 | 99.99 | 99.96 |

*Table 1.  NAS availability.    Source: CAA Annual  Report 1994*
*(time available for use divided by planned time)*

The following example illustrates the sometimes subtle consequences that may follow from lack of data integrity.  NAS provides controllers with printed 'progress strips' on aircraft under control.  The strips show,  among other things, the type of aircraft that is operating the flight.  If this information is in error because for instance the airline has made a change and, for some reason, the change has not been input to NAS, a controller could inadvertently place a fast aircraft behind a slow one without realising it.  If so the gap between the two aircraft could close causing a potential risky loss of separation.   Thus a control decision, perfectly justified on the basis of the data available, could lead to an infringement of separation standards in the airspace. Unavailability—defined as sudden and unexpected loss of the system—is generally less safety significant  because manual recovery is usually possible without endangering safe operation of the air traffic control system.

When NAS was originally designed, some years ago, no specific safety requirements were placed on the design.  So when it was decided to modify NAS the question arose: what extra requirements and criteria, if any, should be placed retrospectively on NAS to ensure safe operation?  And how should any such extra requirements be obtained?   It was clear that the first stage was to identify potential system hazards.  But it was not clear how to go about this.   In process plants the HAZOPS technique has been found to be generally useful for identifying hazards.  This starts with a description of the physical plant and the flows between different elements of the plant and then uses key words to question if hazardous states could arise at any point.  Is such a technique possible with a software system where the flows are data rather than material? Some work at Loughborough University suggested that it is [Broomfield 95], and we therefore explored this approach.  We constructed a simple logical representation of NAS and used key words to question the design and identify potential problems.  The next section describes how this was done in more detail.

# 2.  The Functional Model

The key to any safety analysis is to find a suitable representation for the system being analysed.  In the case of process plants this is straightforward.  In the case of data processing systems however we have potentially many representations to choose from, e.g. Data Flow (Yourdon type) diagrams, State Transition diagrams, Petri nets, etc.  Which to choose?  None of these is very suitable because of the level of detail they introduce.  Secondly none of them combine, in a convenient way, human as well as the machine elements.

At Loughborough they analysed a large number of accidents and incident reports from two different industrial sectors.  They formulated a convenient way to represent the events in terms of a functional model that contained the following generic functions.

* User intervention functions
* Input/output functions
* Communication functions
* Central processing functions

These generic functions or tasks are thought of as making up successive layers of an onion with user interaction functions at the outer layer and central processing functions in the centre.  It turns out that systems such as NAS can be represented quite conveniently and compactly in these terms.  System wide functions are represented as causal threads running through the onion impacting the different layers as necessary.  For instance the production and dissemination of flight progress information is represented as a thread which starts with the manual input of a flight plan (Intervention layer), passes via communication lines (Communication layer) to the database where it is processed and stored (Central Processing layer) and finally via further communication is displayed to the controller (Input/output layer).  The idea is to capture the basic functionality rather than achieve a detailed description suitable say for subsequent design decisions.  Each layer raises different safety issues and it is these that the model attempts to capture.  Once the functions have been identified, a form of HAZOPS can be carried out by applying key words and questions to each function.

In applying this approach to NAS it was found useful first to classify the system data into some fairly broad categories on the basis of safety criticality.  The operational data naturally divided into:

* notification data—data giving notice of impending flight arrival and desired flight trajectory within sector of operation
* situation data—radar derived data indicating current state of each

      aircraft under surveillance

These classes could be further subdivided.  For instance situation data could be divided into:-

* identification data—radar derived identification code
* state data—plan position, height, ...
* derived data—ground speed, ...

These categories clearly differed in their safety criticality, although the NAS design being an integrated design did not recognise any distinction.

To define the requisite system wide functions, each type of data was taken and the processing traced through as a thread from generation or input to termination or output.  Thus 'Provision of Flight Notification Data' was traced through from input of data by manual keying through communication to the central database to output at the required control positions.  The whole function was viewed in terms of threads running through the 'onion' layers.  At each layer these threads typically involved some sub-functions or tasks, e.g. data input or data communication.  The set of all these sub-functions then defined the complete layered functional model for use in the safety analysis.

Our layered functional model differed in emphasis from more conventional ones used in systems analysis and design, e.g  Yourdon type data flow diagrams.  In the latter the  principal concern is to capture system information in way suitable for later design and implementation activities.  Thus data and functions are grouped and merged with a view to convenient design modules, without necessarily retaining relationships with the real world entities represented (less true of object oriented approaches).  What is important in such a description is completeness and detail.  The volume of information usually necessitates a hierarchical grouping within the design description.  For safety analysis, on the other hand, we are interested more in the relationship of the data to the real world than in the structure or grouping of the data within the machine.  If this real world relationship is broken then safety could be endangered, as we have seen earlier.

One problem with the layered approach is that we may end up with too many functions to consider so that the HAZOPS becomes infeasible or uneconomic. To prevent this happening it is necessary to choose the right level of generality to describe the functionality, e.g. treating the amendment of data as the same as its initial input on the basis that in terms of system hazards they are similar if not identical functions. A certain amount of skill is required here. In the case of NAS the description was reduced to twelve high level functions.  It was these functions that formed the basis for hazard identification, analysis and safety requirement generation as described in the following sections. Although there is nothing wrong in principle in having more than twelve functions it is likely in practice that the work involved in subsequent hazard analysis will become overwhelming with many more. Even with the twelve very high level functions defined for NAS, the work of hazard analysis turned out to be quite onerous. It took one analyst, with help from the engineering, operational and safety departments, a period of about nine months to complete.

It is worth noting that the layered functional model approach can be applied at any stage in the system lifecycle. At the requirements stage the model will represent only the basic operational functions foreseen for the system. As the system progresses through its lifecycle further functions may be added to represent implementation considerations, some of which may have been created as a result of previous safety analyses. A good example in the case of NAS was the real time quality control function which vets incoming radar data. This is an important supporting function aimed at ensuring that sensor data, automatically encoded at remote radar stations, is safe to use in the construction of the situation picture. This quality control function would not necessarily have been included at the earliest safety analysis stage. The function uses a variety of geometric, logical, statistical and other consistency checks on the incoming data stream, to report to 'system control' any suspected anomalies. The operator then has the option of deleting the offending radar station as a source of data for further data processing. When the current safety analysis was undertaken, the function already existed and was therefore included and generated hazards of its own when the HAZOPS was undertaken.

A different type of support function, also included, was support for re-sectorisation of the airspace. This task is an essential element of the system and if not available could cause a safety problem for instance at a time when traffic density was increasing and the supervisor required an extra sector to be created in the operations room to cope with the extra workload.

## 3.   Hazard Identification

Hazards were identified by considering the ways in which each high level function could fail. Because each function was 'located' on a specific layer, it had certain general failure properties associated with functions on that layer. This fact could be exploited. For instance any manual input function located at the intervention layer was potentially subject to the typical failures associated with human mistakes. A check list of generic errors could therefore be used when reviewing the functions for failure possibilities.

In fact it turned out that all failures could be considered under the general categories of data/function integrity, or data/function availability. We therefore performed a Failure Modes Effects and Criticality Analysis (FMECA) by applying the following four general headings to each function:

* Loss of data
* Incorrect data
* Loss of function
* Incorrect operation

Applying these headings to the twelve high level functions produced forty eight failure situations, each capable of posing a system hazard. Each failure was discussed by operational staff to determine its probable impact on operations. Some failures were considered not to produce a new hazard. Nevertheless all cases were retained in the FMECA tables for completeness.

| FAILURE MODE | HAZARD |
|---|---|
| Loss of data | Some strips are not output at the required time ahead of flight arrival, probably due to late input of data |
| Incorrect data | Some strip data is erroneous or out of date |
| Loss of function | The output function is not available for a significant period |
| Incorrect operation | Strips printed at wrong position |

*Table 2. Hazard identification table: 'Output Notification Data' function*

As an example of the hazard generation process, the function 'Output Notification Data' produced the system failures shown in table 2, all of which were considered potentially hazardous.

# 4. From Hazards to Requirements

The identified hazards were then used to derive safety requirements for the system in terms of target probabilities required to be achieved by the system in relation to the occurrence of each hazard. We required that the probability of a hazard was as low as possible commensurate with the perceived severity of the hazard's consequences. The first step was therefore to assess the severity of these consequences.

## 4.1 Severity

What do we mean by severity of a hazard and how can we measure it? In principle any of the hazards identified could lead to an accident. But for some hazards the likelihood of an accident is higher than for others. The measure of a given hazard is therefore the likelihood that an accident will occur during the period of time that the hazard in question persists.

To apply this measure directly to the hazards identified in the FMECA would involve a sophisticated and complex probabilistic analysis which would be difficult if not impossible to undertake. It would be necessary to consider the range of events that could take place while the hazardous condition obtained, and then trace the effect onto the actions taken by control staff and via these actions onto the events in the airspace itself. This is a tall order. The calculation would have to take into account not only the options open to the control staff in any given situation, but also the safety margins already built into the control process by way of procedures and separation standards used in different areas of operation. Other variables such as weather, traffic density, and last minute avoidance actions would also have to be taken into account. None of these factors is predictable. It would seem that a large and costly simulation would be needed to evaluate these effects, and even then the results would probably be scenario dependent and therefore open to criticism.

| CAT | EFFECT ON ATC | SITUATION COVERED |
|---|---|---|
| 1 | Inability to provide any degree of ATC for a significant period of time, without warning | Controllers and pilots have no possible means of safely controlling the aircraft and separation will probably be lost |
| 2 | Ability to maintain ATC is severely limited, without warning | Planned separation may not be maintained but contingency measures may be applied to restore the system to a safe state |
| 3 | Ability to maintain ATC is impaired for a significant period of time, without warning | ATC or flight procedures are able to compensate for loss of function but controller/pilot workload may be high thus increasing risk |
| 4 | No effect on the ability to maintain ATC but the situation needs to be reviewed for the requirement to apply some form of contingency measures if the condition persists | There is a lowering of risk, e.g. when a fallback system is lost |

*Table 3. Severity categories*

For the above reasons the approach taken by National Air Traffic Services Ltd is based on assessing the effect on the ability to control, rather than the effect on the airspace situation itself. This approach is similar to that employed in other industry sectors, e.g. in the automotive industry safety is assessed in terms of the effect on the ability of the driver to control his vehicle rather than any consequent damage or loss of life, effects which are difficult to estimate with confidence.

Following this approach, hazards to ATC are assigned to one of four severity categories, as shown in table 3. Note the 'without warning' caveat attached to the definitions. It is only when there is no warning that a hazard arises because otherwise the ATC system can preserve safety by reducing the volume of traffic to be handled.

Using these definitions we assigned a severity category to each NAS failure identified as hazardous. To assist the process we considered both the overall effect on the system and any arrangements already in place that might mitigate the effects  For example in the case of the total loss of the function that outputs Notification Data we considered the aspects shown in table 4.   Based in this type of analysis a judgement was made regarding the severity of each identified hazard.   This was one of the most difficult parts of the work because it depended on getting an agreed view among engineers, operational staff and safety professionals as to the seriousness of hypothetical situations.

| ASPECT | DESCRIPTION |
|---|---|
| Overall effect | Controllers are not provided with warning strips; current strips are not updated |
| Compensatory provisions | Flight plan data is printed out periodically so that hardcopy is available in the event of system failure |
| Recovery actions | If the outage is lengthy, the ATC operation reverts to 'manual' ; traffic flow is reduced; progress strips  are produced by hand from backup hardcopy |

*Table 4.  Effects and mitigations: Failure to Output Notification data*
*(part of  the FMECA)*

Most of the identified hazards were judged to fall into category 4, with a few at category 3.  No hazard was judged to be more severe than category 3.  In some cases it was felt that there was no hazard at all.


## 4.2  Risk and Probability Targets

Four levels of risk are defined.  The most serious risk level is A which is unacceptable in any circumstance.  The next level, B is undesirable, but acceptable exceptionally. The third level C is acceptable with reservations and D the lowest risk level is acceptable without reservations. The National Air Traffic Services Ltd system requires management approval before systems can be accepted at any level of risk. For the higher levels more senior management must sign.

Systems are designed to achieve the lowest level of risk that is reasonably practicable, on the ALARP principle—as low as reasonably practicable. A Risk Tolerability Table is used to indicate the level of probability that the system should aim to achieve, given the severity of the hazard that the system is judged to present. Probabilities are 'quantised' into 6 ranges as shown in Table 5.

| CLASS | MEANING | RANGE |
|-------|---------|-------|
| Frequent | Many times in system lifetime | $P_s > 10^{-3}$ |
| Probable | Several times in system lifetime | $P_s = 10^{-3}$ to $10^{-4}$ |
| Occasional | Once in system lifetime | $P_s = 10^{-4}$ to $10^{-5}$ |
| Remote | Unlikely in system lifetime | $P_s = 10^{-5}$ to $10^{-6}$ |
| Improbable | Very unlikely to occur | $P_s = 10^{-6}$ to $10^{-7}$ |
| Extremely improbable | Incredible | $P_s < 10^{-7}$ |

*Table 5. Probability ranges*
*(per operational hour)*

The target probabilities obtained in this way were generally in the range of 'Probable', which means between 1 in $10^{-3}$ and 1 in $10^{-4}$ occurrences per operational hour in a sector of operations. This is quite a tough target to meet, especially for the intervention functions where the possibility of operator mistakes has to be allowed for.

These target probabilities, one for each identified hazard, were taken as the safety requirements of the system.

# 5. Concluding remarks

We have described how a layered functional model was used to assist the process of identifying hazards, and then how these hazards were used to generate safety requirements for the system. We found that the layered functional model provided a convenient and appropriate way of describing the system for safety analysis purposes. The model combines hardware, software and human processes, all of which are possible sources of failure, in a logical and convenient way. If this approach were to be adopted more generally, a more uniform treatment might be possible within a given company or industry. A standard checklist of considerations could be compiled for each layer. For instance at the intervention layer many of the problems arise due to the possibility of human error. A check list of potential safety protective measures could be employed to assist in the task of formulating appropriate counter-measures.

This raises the interesting possibility that it might be possible to automate part of the process, for instance by providing tools to assist the process of hazard identification. The advantage would be that the checklists could be added to and refined over time and applied consistently across a given organisation. There might also be gains in productivity, allowing more functions to be considered.

Finally some thoughts on the difficulties of applying the quantified risk approach. The probabilities that we end up with are usually very small, as can be seen from table 5. For avionics systems even lower target figures are common—as low as $10^{-9}$ per flight hour for instance [JAR-25]. There is considerable difficulty in verifying that such target figures have been met. Such low probabilities usually cannot be verified directly, for example by running the system and checking that it does not fail after the requisite number of hours. Again, it is usually not possible to predict the failure rate using any form of statistical model, as is done for instance in the hardware field where components are subject to random wearout failures at well established failure rates. With a new system containing significant software most of the failures are systematic rather than random, and corrected as soon as found. Moreover there is usually no system history to indicate the basic failure rates. Any predictive statistical model of performance is therefore of dubious validity.

If these probability targets cannot easily be verified, either directly or indirectly, how are we to interpret them? And how can a system be accepted into service with confidence that it has met its safety requirements?

In the software field it now seems to be fairly well accepted that the best approach is to concentrate attention not on the software itself but on the process of developing the software. Thus different levels of reliability are interpreted as implying more or less rigour in the development process. The emerging software engineering standards also include recommended design features, e.g. for failure recovery. This prompts the thought that perhaps at the system level too reliability requirements should be posed more in terms of design philosophies and policies, as is done in the security field, rather than in terms of quantitative performance targets. This thought, if followed, would lead to a more prescriptive approach to design for safety, and one that could perhaps be more easily defended.

## Acknowledgements

The author would like to thank the Directors of National Air Traffic Services Ltd for permission to publish this paper. The author would also like to acknowledge the work carried out at Loughborough University under the AUSDA project, which provided the basis for the approach. The author would also like to thank Mr M H Davies for useful comments on the initial draft. The views expressed in this paper are the author's own and do not necessarily reflect the views of National Air Traffic Services Ltd.

## Glossary

| | |
|---|---|
| ALARP | As Low As Reasonably Practicable |
| ATC | Air Traffic Control |
| FMECA | Failure Modes, Effects and Criticality Analysis |
| HAZOPS | Hazards in Operations |
| NAS | National Airspace System |

## References


[Leveson 95]   Leveson N. Safeware:  System Safety and Computers, Addison-Wesley, N.Y., 1995

[Broomfield 95]   Broomfield E.J. and Chung P.W.H.  Using Incident Analysis to Derive a Methodology for Assessing Safety in Programmable Systems, in Achievement and Assurance of Safety, Redmill F. and Anderson P., editors, Spinger-Verlag, 1995

[JAR-25]   Joint Airworthiness Requirement 25, Joint Aviation Authorities, Brussels, 1990

# Formal Methods:
# No Cure for Faulty Reasoning


Martin Loomes and Rick Vinter

Faculty of Information Sciences, University of Hertfordshire

Hatfield, United Kingdom



## Abstract

Owing to the benefits commonly associated with their use and links with
scientific culture, formal methods have become closely identified with the
design of safety-critical systems. But, despite the mathematical nature
of the logic systems underlying most formal notations, many aspects of
formal methods are much less predictable than one might realise. Specifically, it is suggested that the ways in which people interpret and reason
about formal descriptions can lead to similar kinds of errors and biases as
those exhibited during previous cognitive studies of logical statements in
natural language. This paper reports a series of preliminary experiments
aimed at testing this hypothesis and several related issues. Early results suggest that, in reality, people frequently depart from fundamental
principles of mathematical logic when reasoning about formal specifications, and are content to rely upon probablistic, heuristic methods. Furthermore, they suggest that manipulating such factors as the degrees of
thematic and believable content in formal specifications can lead to significant reasoning performance enhancement or degradation. So, although
faulty reasoning cannot be cured by formalisation alone, it would appear
that the human potential for error can be reduced by avoiding certain
expressions and choosing alternative, equivalent forms.


# Introduction

One of the strategies commonly adopted in the design of safety-critical systems
is to imagine what can possibly go wrong and to consider ways of preventing or
containing the errors. At the technological level we are quite adept at predicting
and containing the consequences of failure of the system itself, through a variety
of tried and tested techniques. We are less experienced, however, at predicting
and avoiding the human errors in the system design process that lead to these
failures. For example, the report on the Ariane 5 disaster [Lions96] clearly
attributes the failure to faulty design which, with the benefit of hindsight, is
simple to understand. However, whilst the recommendations in this report are
specific at the technical level in identifying how the design fault could have been
found during testing, they are rather vague in terms of improving the design
process to avoid such errors arising in future. There are some global approaches
that have gained support in recent years, such as adopting particular design
methods or documentation standards, but it is far from clear that these really
deliver much beyond a certain feel-good factor that comes from doing something
rather than nothing.

A particular area that has become identified with safety-critical design is that of formal methods. Indeed, many designers who reject formal methods for their own use on commercial or industrial projects will often add the caveat that they can see their virtue for safety-critical systems. Why is this the case? Partly one suspects it is because formal methods, and mathematics in general, are associated with a scientific culture [Hoare84], and this is the culture that we, as consumers, hope and believe underpins professions such as medicine and engineering where new developments are welcomed but acknowledged as potentially life-threatening. "The appliance of science"[1] conjures up visions of careful, well thought-out and thoroughly analysed technological progress, based on mathematical formalisation and reasoning. Indeed, many advocates of formal methods in the past two decades have drawn extensively on analogies with other technical disciplines to demonstrate how we should seek these scientific and mathematical foundations for software if we aspire to becoming a professional engineering discipline.

Apart from the generally perceived, if nebulous, advantages of working within a scientific culture, adopting formal methods is thought to bring a number of specific benefits. Anyone who asks an undergraduate to list the advantages of formal methods is likely to get an answer containing words such as "unambiguous", "precise" and "correct". This can rapidly become interpreted as implying that the use of formal methods leads to descriptions of a system that are only open to one interpretation, mean exactly what we want them to mean, and where we can show that the system has exactly the properties we require. In fact, of course, it is only the syntax (and possibly a formal semantic interpretation) that is unambiguous and precise. How a reader chooses to interpret the description into some real-world problem domain, reason about the system across this interface and act upon the conclusions is rather less predictable or controllable.

In formalising computer-based systems, we are usually seeking to automate information processing in some form or another, and what needs to be captured is the reasoning process itself, where the obvious models come from logic, which does not currently enjoy the same status and shared culture as the mathematics traditionally applied in other realms of engineering. Logic, which was once a mainstream curriculum topic, is now distributed between mathematics, philosophy and linguistics, making it a complex beast to study and pin down. Thus whilst it may be very sensible to assume that two civil engineers reading a set of equations governing fluid flow will interpret them in the same way and reason about them to the same conclusions, it is far from obvious that this desirable behaviour will necessarily carry over to two software engineers reading formal specifications.

---

[1]This phrase has become well known in the UK as an advertising slogan for Zanussi washing machines.

"It (logic) is justified in abstracting - indeed it is under obligation to do so - from all objects of knowledge and their differences, leaving the understanding nothing to deal with save itself and form"    Kant [Smith93, p. 18].

Proponents of formal methods often adopt a point of view similar to Kant's, which suggests that formal, abstract, reasoning will be fault-free (save for possible slips which are unlikely to be replicated during subsequent analysis and thus will be easily spotted) as it is liberated from distractions such as intuitions and background knowledge. Whilst this view may be defended from a theoretical perspective, defining formal reasoning as perfect and everything that deviates from it as erroneous, the pragmatics of the situation become rather different. There have been several studies that show actual reasoning performance improves as the task becomes less abstract [Dominowski95, Griggs82, VanDuyne74, Wason71, Wilkins28]. Moreover, there are errors that are made systematically by large numbers of people which are unlikely to be spotted by naive inspection, uninformed of the likely sources of errors.

The COPSE project [Loomes94] was established to explore some of the cognitive and organisational factors that influence software engineers. Whilst most of the work has focused on the organisational and cultural issues, work has also started on the analysis of cognitive issues in the use of formal methods. There are clearly a number of possible starting points for such an investigation, and a number of places where emphasis could be placed. For example, studies of the differences between the use of various notations for formalisation, or differences between problem domains, utilising case studies or demonstrator projects, might yield significant results. The difficulty posed by this sort of high-level case study approach is that there are usually so many factors involved that it proves very hard to devise repeatable experiments, or to explain the results in terms of plausible theories that lead on to practical new experiments. Too often such studies tend to lead to anecdotal evidence, aimed at defending a favoured hypothesis rather than exposing a scientific hypothesis to scrutiny.

One current strand of work at Hertfordshire is attempting to pose and answer, using empirical techniques, a well-founded set of questions based on existing theoretical bodies of knowledge surrounding these issues. In order to achieve this, considerable refinement of the issues has been undertaken to reduce the number of factors under consideration at any one time; hopefully, this has been done without naive over-simplification which would render the results too far removed from real engineering practice. First, we are concentrating primarily on the interpretation of existing specifications, rather than the creative processes that lead to new specifications. Second, we are using a basic set of logical tools for most specifications, rather than an enhanced mathematical tool-set involving structures such as lists, functions and relations. Finally we are using the concrete syntax provided by the Z specification notation [Spivey92]. Although this introduces possible confounding factors by the use of a schema notation, Z is sufficiently popular to ensure that knowledgeable users can be found as participants for the experiments. Within this framework, a number of studies are underway to explore systematic errors in the interpretation of Z specifications and subsequent reasoning errors.

In order to ensure that these studies are based on existing bodies of knowledge, the starting point has been the psychological literature on logical reasoning. There have been many studies carried out over the years in which hard (that is, scientifically repeatable) results have been achieved showing that certain forms of logical expression can lead to faulty reasoning [Braine91, Johnson-Laird72, Lakoff71, Newstead83]. Most of these studies have been carried out using problems posed in natural language, with participants drawn from the general public. The initial question this project set out to explore was whether these results carry over to the realm of software engineers using an established formal notation which they believe they understand. If so, can we use the findings to make available tools and techniques which will aid designers in identifying areas of formal descriptions that are "at risk", and where defensive approaches need to be taken, perhaps by associating metrics with particular forms, or even banning the use of certain syntactic constructs in the specification of safety-critical systems? In this way we hope to develop a technological understanding of formal specification languages and their use to mirror the developing understanding of programming.

## The Experiments

A number of areas of potential interest have been identified by analysis of the psychological literature which intersect with reasoning tasks commonly found in software engineering. Examples of these include: the problems of reasoning with implications, the tendency of readers to guess what formal text means based on intuitive interpretation regardless of any formal semantics, preferred styles of expression, the problems of disjunctive and conjunctive reasoning, and syllogistic reasoning with quantification. Space does not permit a detailed discussion of all the experiments and results to date, but this section gives an overview of the approach and highlights a few of the findings which might cause us to reflect on some of the received wisdom concerning the use of formal methods, including the degree of safety we associate with their use. The experiments are loosely clustered into two areas, those concerned with syntactic structure and its interaction with thematic content and those concerned with other features of the specifications under consideration such as their believability and literary style. As the project progresses these two areas will be brought together in more complex experiments exploring the interaction.

Two types of experiment are discussed below: pilot experiments, which were undertaken primarily to help refine the questions and methodology, and the main experiments which constitute the substantive part of the project. The pilot experiments were conducted on small numbers of participants and no claim is made for the statistical significance of their results, although they do suggest some interesting areas for further study. The main experiments are currently being conducted on far larger groups and are intended to provide statistically significant results in the suggested areas. Some of these are nearing completion and tentative results are mentioned.

# Syntactic Features

A particularly famous study of human reasoning is the Wason four card problem [Wason66]. Subjects are confronted with a problem similar to that that shown in Figure 1. In abstract, logical, terms the rule is of the form $p \Rightarrow q$, and the four cards represent instances of $p$, $q$, $\neg p$ and $\neg q$. The "correct" cards to turn over are A ($p$) and 7 ($\neg q$), as these are the only instances that can falsify the rule conclusively. Turning over the 4 ($q$) card may increase our confidence by supplying positive evidence for the rule, but it will not help us to test it. Wason found, and this is a fairly repeatable result which one of the authors has regularly replicated with large groups of students studying logic, that although virtually every participant correctly selects the $p$ case as relevant very few select the $\neg q$ case. Moreover, it is quite common to select both $p$ and $q$, thus missing one test and carrying out an unnecessary one.

A pack of cards has letters on one side and numbers on the other. Here is a rule: "If there is an A on one side of the card then there is a 4 on the other". Here are four cards from the pack lying on a table.

| A | S | 4 | 7 |

Which card(s) would you need to turn over in order to establish whether the rule is true or false?

Figure 1: Wason's abstract selection task.

One of the claims sometimes made for formal methods is that they help the process of test-set generation. Moreover, we might expect that if we make the implication explicit, by expressing it formally, we would cue the reader into potential problems, especially since these were almost certainly discussed when the notation was first taught. With this in mind, a logically equivalent problem to Wason's task was posed in Z (Figure 2) and carried out by a number of computer scientists with differing levels of Z experience. The aim was to see if there was any substance to this claim, and whether Wason's results would carry over into a formal expression of the task.

The requirements for software operation *InOut* are: "If the operation receives an A as input then it will output a 4". Its formal specification follows.

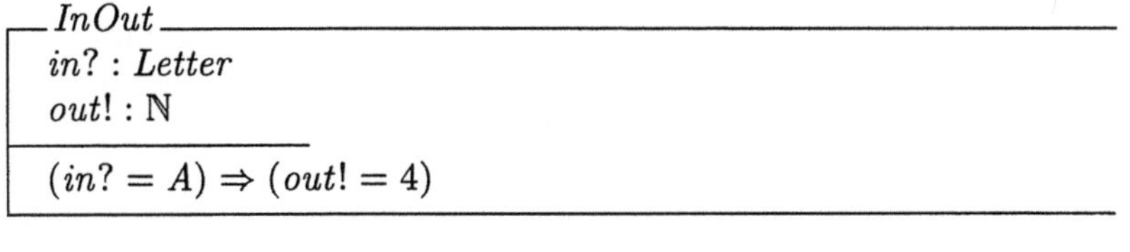

Which inputs and outputs would enable you to test whether *InOut* is working correctly, according to its requirements?

Figure 2: The formalised selection task.

In spite of all the cues given, and the fact that participants were given unlimited time to complete the task, their performance on this experiment was actually worse than that of the "man in the street" on Wason's task, although only marginally so. Generally, a very close correlation between the results of this experiment and Wason's results was noted. No participant correctly recognised the significance of the $\neg q$ case - as compared with 4% in Wason's experiment [Wason72, p. 182]. Every participant correctly identified the $p$ case as necessary, but (as in Wason's experiment) the most popular choice of combination was $p$ and $q$. One possible explanation of this phenomenon is offered by Evans [Evans72], who suggests that in this sort of reasoning task people are often guilty of a "matching bias", preferring to give answers that contain the same terms as are contained in the problem presentation. In this case, both A ($p$) and 4 ($q$) appear in the question so they are preferred terms in the solution. It is important not to read too much into a very simple pilot experiment of this nature, although the results do suggest that we do not necessarily achieve improved reasoning performance by formalisation alone, and we ought to be aware of the possibility of matching bias when we subject formal specifications to tests such as walk-throughs.

One major criticism of the Wason task is that the task is highly abstract: no-one can really imagine why we might have cards of this type in existence. Given the following problem, which is formally identical but has thematic content, people rarely make mistakes.

> Here is a rule: "If a person is drinking alcohol, then the person must be over 18 years of age". There are four young people drinking in the bar and we know just one fact about each: one is drinking a beer, one is drinking lemonade, one is 15 years of age, and one is 20 years of age. Which of the youths would you need to question in order to establish whether the rule is being conformed to?

This suggests that we should explore not only formalisation, but also the degree of abstraction away from thematic content as we formalise. In order to pursue this, a set of experiments has been devised to evaluate reasoning performance in situations with varying degrees of formality and thematic content. Three groups of participants are asked to complete three different types of task. In the first group an abstract formal task is set, and participants reason formally about shapes and colours, with no obvious thematic connection. In the second group a thematic formal task is set, and participants deal with formally presented situations such as the relationship between the safety status of a nuclear reactor and the temperature of its coolant. The third group are set the same tasks as the abstract formal group, but with the tasks being presented in natural language. In all three groups, participants are given a series of questions which comprise a statement about the system involving implication, together with a premise about the system state. Rather than generating test cases as in the Wason experiment, the participants are asked to draw conclusions from the given information or to state that no conclusion is possible, and are also asked to provide confidence ratings on their answers. By manipulating the forms of the presented implication and the premises, we can elicit details of reasoning performance corresponding to different types of logical inference (*modus ponens* and *modus tollens*) and fallacious reasoning (*denial of the antecedent* and *affirmation of the consequent*) for all positive and negative combinations of the premises. For example, Figures 3-5 show one of the tasks set to explore the affirmation inference for all three groups. Formally, no valid conclusion can be drawn, but the psychological literature suggests that participants will fallaciously use *affirmation of the consequent* to draw conclusions.

If $colour' \neq blue$ after its execution, what can you say about the value of $shape$ before operation $SetColour$ has executed?

$$\begin{array}{|l|}
\hline
__\ SetColour \ ____________________ \\
\quad \Delta ShapeAndColour \\
\hline
\quad (shape = circle) \Rightarrow (colour' \neq blue) \\
\quad shape' = shape \\
\hline
\end{array}$$

(A)  $shape \neq rectangle$      (C)  $shape \neq circle$
(B)  $shape = circle$      (D)  Nothing

Figure 3: An abstract formal logic based task.

If $\neg(reactor\_status! = Ok)$ after its execution, what can you say about $coolertemp$ before operation $ReactorTempCheck$ has executed?

$$\begin{array}{|l|}
\hline
__\ ReactorTempCheck \ _________________ \\
\quad \Xi NuclearPlantStatus \\
\quad reactor_status! : Report \\
\hline
\quad coolertemp > Maxtemp \Rightarrow \neg(reactor_status! = Ok) \\
\hline
\end{array}$$

(A)  $coolertemp \leqslant Maxtemp$      (C)  $coolertemp > Mintemp$
(B)  $coolertemp > Maxtemp$      (D)  Nothing

Figure 4: A thematic formal logic based task.

If the shape is a circle then the colour is not blue.
The colour is not blue.

Based on the above description, what can you say about shape?

(A)  The shape is not a rectangle      (C)  The shape is not a circle
(B)  The shape is a circle      (D)  Nothing

Figure 5: An abstract natural language based task.

This experiment is still ongoing, and so definitive results cannot be given, but tentative analysis of the data collected so far suggests that many of the results previously observed in the psychological literature carry across to experiments where the tasks are expressed in Z, and the participants are all software engineers trained in formal methods. It also suggests that there are some differences between the groups. For example, as expected, all three groups have very little difficulty with *modus ponens* reasoning. *Modus tollens* reasoning seems to be performed better in the formal groups than in the natural language group and the thematic formal group seems less prone to fallaciously *denying the antecedent*. The natural language group, however, seems less prone to *affirming the consequent*. In general, the formal thematic group seems to be performing slightly better than the other groups, and also seems to have more confidence in the answers given. If these results are born out in the final analysis it would suggest that formalisation can slightly improve reasoning performance with implications, but that certain forms of expression should be avoided and equivalent forms chosen. We should note, however, that even in the performance of the best group, several examples of faulty reasoning occur and the increased confidence of the group could lead to less rigorous testing which offsets any potential benefits.

Similar experiments involving abstract and thematic groups are being carried out to explore reasoning performance with conjunctive and disjunctive forms in Z: an area less well represented in the psychology literature. In particular, some of the contextual dependencies involved in the use of inclusive and exclusive disjunctions are being varied to see if these lead to differences in performance. The literature suggests that people generally find it easier to reason with exclusive forms, but formal logical systems frequently omit this from the primitive syntax, perhaps for reasons of theoretical elegance. These experiments are attempting to isolate features of individual connectives, together with positive and negative instances of propositions. Once reliable results from these experiments are available, further experiments will be carried out with compound forms, involving all combinations of connectives, and also quantifiers will be introduced. There is a significant amount of prior work on faulty reasoning with quantifiers, and the major thrust of this work will be to see if these results carry across to our problem domain, or whether improvements in performance come about with formalisation in Z.

# Evaluation and Interpretation of Specifications

Several experiments have been initiated aimed at identifying the cognitive processes involved when readers are asked to evaluate the quality of given specifications or translate them into natural language. In particular, two pilot experiments have been carried out to explore how effective the use of formal specifications is at communicating system concepts, and how engineers themselves assess the quality of the specifications. In the first experiment, for example, we set out to test Gravell's assertion, based on an informal "straw poll" of software engineers opinions, that "To communicate clearly with the majority of readers you should, in general, prefer clarity to brevity" [Gravell90, p. 139]. A group of software engineers were presented with an English description of a simple operation "Toggle", which exchanges the current state of a simple two-way switch, and the following four formal descriptions of the same system.

$$\begin{array}{|l|}\hline \text{\_\_}\textit{Toggle}\text{\_\_\_\_\_\_\_\_\_\_\_\_\_\_\_\_\_} \\ s, s' : SWITCH \\ \hline s' \neq s \\ \hline \end{array}$$

Concise

$$\begin{array}{|l|}\hline \text{\_\_}\textit{Toggle}\text{\_\_\_\_\_\_\_\_\_\_\_\_\_\_\_\_\_} \\ s, s' : SWITCH \\ \hline (s = \textit{off} \land s' = \textit{on}) \lor \\ (s = \textit{on} \land s' = \textit{off}) \\ \hline \end{array}$$

Verbose

$$\begin{array}{|l|}\hline \text{\_\_}\textit{Toggle}\text{\_\_\_\_\_\_\_\_\_\_\_\_\_\_\_\_\_} \\ s, s' : SWITCH \\ \hline s = \textit{on} \Rightarrow s' = \textit{off} \\ s = \textit{off} \Rightarrow s' = \textit{on} \\ \hline \end{array}$$

Precise

$$\begin{array}{|l|}\hline \text{\_\_}\textit{Toggle}\text{\_\_\_\_\_\_\_\_\_\_\_\_\_\_\_\_\_} \\ s, s' : SWITCH \\ \hline (s = \textit{on} \lor s = \textit{off}) \Rightarrow \\ (s' = \textit{on} \lor s' = \textit{off}) \\ \hline \end{array}$$

Imprecise

Figure 6: The four styles of formal specification.

These four formal specifications could be classified as concise, precise, verbose and imprecise. The imprecise version under-specifies the system, but the other three are formally equivalent. The participants were asked which version best described the system's behaviour, and to justify their choices. They all rejected the imprecise version, but were split evenly between the other three versions. There was an interesting correlation between age, experience and chosen style, with the older, more experienced engineers preferring the precise style, and the younger, less experienced engineers preferring the concise style. There are many factors that could explain this, including prior experience of specific problems, educational backgrounds or cultural trends. What is significant, however, is that in any project team there are likely to be engineers working with different preferred styles, frequently being asked to work with presentations not in their preferred style. This also highlights the fact that we cannot ignore the individual differences between engineers when attempting to identify "good practice" in design or devise training programs for their development. One possible avenue of exploration is to investigate subsequent reasoning performance based on precise, concise and verbose forms to see if there are any significant differences.

More worrying is the result of another pilot study where participants were shown the specification of a system containing a counter-intuitive clause. They were asked to translate a given schema into natural language in order to test the claim made by Liskov and Berzins that "there is only one way to interpret a formal specification because of the well defined and unambiguous semantics of the specification language" [Liskov79, p. 279]. Whilst it may be true that there is only one formal interpretation of the given schema into some abstract denotational semantics, what is of practical interest is whether real software engineers will reflect this single interpretation in their own understanding of the specification, and the way their interpretation governs their behaviour.

$$
\begin{array}{|l}
\hline
\_\,Library \,\_\_\_\_\_\_\_\_\_\_\_\_\_\_\_\_\_\_\_\_\_\_\_\_\_\_\_\_\_\_\_\_\_\_\_ \\
\quad stock : Copy \twoheadrightarrow Book \\
\quad issued : Copy \twoheadrightarrow Reader \\
\quad shelved : \mathbb{F}\; Copy \\
\quad readers : \mathbb{F}\; Reader \\
\hline
\quad shelved \cup \mathrm{dom}\; issued = \mathrm{dom}\; stock \\
\quad shelved \cap \mathrm{dom}\; issued = \varnothing \\
\quad \mathrm{ran}\; issued \subseteq readers \\
\quad \neg\,\exists\, r : readers \bullet \neg(\#(issued \rhd \{r\}) > maxloans) \\
\hline
\end{array}
$$

Original fourth predicate: $\forall\, r : readers \bullet \#(issued \rhd \{r\}) \leq maxloans$
The number of books that any reader borrows must be less than or equal to the maximum number of loans allowed.

Revised fourth predicate: $\neg\,\exists\, r : readers \bullet \neg(\#(issued \rhd \{r\}) > maxloans)$
The number of books that any reader borrows must be more than the maximum number of loans allowed.

Figure 7: The library specification - modified from [Potter91, p.124].

Most engineers gave correct interpretations of the first three clauses, which were consistent with their intuitions, although we should perhaps be concerned with the 25-33% who erred in each case! However, all of the participants failed to provide a correct interpretation of the final clause. In each case, the participant provided an interpretation which was consistent with an intuitive understanding of library systems, rather than the text as given. The semantics certainly appeared to be unambiguous to the readers, but the meaning was not what the author intended. One possible explanation of this is that readers use the formal specification to obtain linguistic cues regarding the domain of interest, then fit the mentioned terms into relations that are consistent with their intuitions. If this is the case we should be deeply concerned, as it suggests that far from assisting the reasoning process by liberating the engineer from errors caused by faulty intuitive reasoning, we may in fact be obscuring intuitive reasoning by formalisation: achieving a feel-good factor with "the appliance of science" that is not deserved. The fact that every participant made the same mistake suggests that this sort of error is unlikely to be picked up by testing. Clearly this was an extreme example, deliberately chosen to be directly counter-intuitive, but the strength of the result suggests we should take it seriously.

Indeed, there are a number of examples in the psychology literature that suggest people generally tend to abandon logical principles for reasoning in favour of heuristic and probabilistic methods when confronted with arguments containing information relating to strongly-held beliefs [Janis43, Morgan44, Evans83, Oakhill90]. The experiments carried out so far suggest that this phenomenon carries across to our problem domain. A further set of experiments is currently being carried out to explore this systematically. In addition to the two dimensions of formal/informal and abstract/thematic, believable/incredible will be added. Although we are not really interested in the extreme of incredible, eliciting the belief structures and strength of opinion from engineers may provide pointers to potential sources of reasoning errors.

# Conclusions

The results of experiments carried out so far, although still tentative, indicate that many of the errors in reasoning that have been noted by psychologists in experiments with ordinary people, working in a natural language, seem to arise just as frequently with software engineers working in Z. Although some improvements have been observed in a few specific situations, the improvements are not as dramatic as one might expect from reading some of the more evangelical literature from proponents of formal methods. There are at least two possible reactions to this. We might start to doubt that formal methods have a role to play in the design of safety-critical systems at all. Perhaps more constructively, however, we might see this as an opportunity to explore how reasoning errors arise, and to develop ways of working that defend against them.

One of the strengths of formalisation in this context is that the grammatical structures are well-defined, and hence we can carry out well-controlled experiments in ways that we cannot easily do with natural languages. Moreover, we can imagine tools that might highlight potential areas of concern in formal specifications, and suggest alternative equivalent logical forms that are less prone to causing errors. We might even learn some lessons from the use of formal methods that we can carry across into less formal reasoning about systems. The main aim of this paper, however, is not to influence the perceptions or use of formal methods. The results are still far too sketchy to warrant this. Rather, the authors' hope is that we will open up the systematic analysis of these sort of issues as a topic of research in Software Engineering. In our opinion, it is dangerous to expend all our effort on the developments of methods and notations in the discipline based on anecdotal evidence or case study material that cannot be easily replicated or generalised.

# References


[Braine91] Braine M.D.S. and O'Brien D.P., A theory of If: A lexical entry, reasoning program, and pragmatic principles. *Psychological Review, 98*, 182-203, 1991.

[Dominowski95] Dominowski R.L., Content effects in Wason's selection task. In S.E. Newstead and J.St.B. Evans (Eds.), *Perspectives on Thinking and Reasoning. Essays in Honour of Peter Wason.* Hove UK: Lawrence Erlbaum Associates, 1995.

[Evans72] Evans J.St.B.T., Interpretation and matching bias in a reasoning task. *Quarterly Journal of Experimental Psychology, 24*, 193-199, 1972.

[Evans83] Evans J.St.B.T., Barston J.L. and Pollard P., On the conflict between logic and belief in syllogistic reasoning. *Memory and Cognition, 11* (3), 295-306, 1983.

[Gravell90] Gravell A., What is a good formal specification? In J.E. Nicholls (Ed.), *Z User Workshop, Oxford 1990. Proceedings of the Fifth Annual Z User Meeting, Oxford 17-18 December 1990*, Springer-Verlag, 1990.

[Griggs82] Griggs R.A. and Cox J.R., The elusive thematic materials effect in the Wason selection task. *British Journal of Psychology, 73*, 407-420, 1982.

[Hoare84] Hoare C.A.R., Programming: Sorcery or science, *IEEE Software*, 5-16, April 1984.

[Janis43] Janis L. and Frick F., The relationship between attitudes toward conclusions and errors in judging logical validity of syllogisms. *Journal of Experimental Psychology, 33*, 73-77, 1943.

[Johnson-Laird72] Johnson-Laird P.N. and Tridgell J.M., When negation is easier than affirmation. *Quarterly Journal of Experimental Psychology, 24*, 87-91, 1972.

[Lakoff71] Lakoff R., If's, and's, and but's about conjunction. In C.J. Fillmore and D.T. Langendoen (Eds.), *Studies in Linguistic Semantics.* New York: Holt, Rinehart and Winston, 1971.

[Lions96] Lions J.L., *Ariane 5: Flight 501 Failure. Report by the Inquiry Board.* Paris: European Space Agency, 19 July 1996.

[Liskov79] Liskov B. and Berzins V., An appraisal of program specifications. In P. Wegner (Ed.), *Research Directions in Software Technology*, Cambridge, Mass: MIT Press, 1979.

[Loomes94] Loomes M., Ridley D. and Kornbrot D.E., Cognitive and organisational aspects of design. In F. Redmill and T. Anderson (Eds.), *Technology and Assessment of Safety-critical Systems*, Springer-Verlag, 1994.

[Morgan44] Morgan J.J.B. and Morton J.T., The distortion of syllogistic reasoning produced by personal convictions. *Journal of Social Psychology, 20*, 39-59, 1944.

[Newstead83] Newstead S.E. and Griggs R.A, The language and thought of disjunction. In J.St.B.T.Evans (Ed.), *Thinking and Reasoning. Psychological Approaches*, 76-106, London: Routledge and Kegan Paul, 1983.

[Oakhill90] Oakhill J., Garnham A., and Johnson-Laird P.N., Belief bias effects in syllogistic reasoning. In K.J.Gilhooly, M.T.G. Keane, R.H. Logie, and G. Erdos, *Lines of Thinking: Reflections on the Psychology of Thought. Volume 1. Representation, Reasoning, Analogy and Decision Making*, 125-138, Chichester: John Wiley and Sons, 1990.

[Potter91] Potter B., Sinclair J. and Till D., *An Introduction to Formal Specification and Z.* Hemel Hempstead: Prentice-Hall, 1991.

[Smith93] Smith N.K. (translator), *Immanuel Kant's Critique of Pure Reason.* Second Edition, London: Macmillan, 1993.

[Spivey92] Spivey J.M., *The Z Notation: A Reference Manual.* Second Edition. Hemel Hempstead, Prentice Hall International, 1992.

[VanDuyne74] Van Duyne P.C., Realism and linguistic complexity in reasoning. *British Journal of Psychology*, 65, 59-67, 1974.

[Wason66] Wason P.C., Reasoning. In B.M. Foss (Ed.), *New Horizons in Psychology. Volume 1*, Reading: Penguin, 1966.

[Wason71] Wason P.C. and Shapiro D., Natural and contrived experience in a reasoning problem. *Quarterly Journal of Experimental Psychology, 23*, 63-71, 1971.

[Wason72] Wason P.C. and Johnson-Laird P.N., *Psychology of Reasoning: Structure and Content.* London: Batsford, 1972.

[Wilkins28] Wilkins M.C., The effect of changed material on ability to do formal syllogistic reasoning. *Journal of Social Psychology, 24*, 149-175, 1928.

# Artificial Intelligence - Genuine Hazards?

Ken Frith and Richard Ellis
Crew Services Ltd
Portsmouth, Hampshire, UK

### Abstract

System designers are increasingly faced with pressures to utilise intelligent systems as a part of their design. These pressures can arise from the functional requirement, which only intelligent systems can meet; from the customer, demanding the use of 'fashionable' technology; or from the designer's own wish to extend the capabilities of his design. Although the safety aspects of lower orders of intelligent system can still be managed satisfactorily by current analysis techniques, the authors feel that the increasing intelligence of these systems requires a significant departure from traditional methods. This discussion paper examines some of the problems of introducing intelligence into safety-related systems, and explores possible alternatives to traditional methods of making safety cases for such systems.

## 1.    Introduction

The authors, as systems engineers with responsibilities for analysis of systems during high-level design, are customarily involved with projects at the requirements, concept and feasibility stages. This has two conflicting facets: requirements and concept design generates ambitious, often fanciful, solutions; feasibility study introduces caution, and tends to highlight risks such as cost, novel technology and, of course, safety. Among the novel technologies, the use of Artificial Intelligence (AI) in safety-related systems has been growing apace in recent years, yet it seems that there is a scarcity of clear information and guidance on its impact on safety.

The pressure for system designers to introduce intelligence into their systems is high, and a safety engineer is increasingly exposed, often being inadequately armed with the knowledge, guidance or techniques needed to give proper advice and adjudication on the subject. This paper therefore begins by explaining what we mean by 'Artificial Intelligence', in particular with respect to its impact on safety-related systems. From this, we aim to differentiate between AI that is manageable by current analysis techniques, and that which presents the safety engineer with real problems. This is an inquisitive paper, and not aimed at offering solutions; its purpose is to ask questions and to outline and try to understand some of the problems that arise from the increased use of AI in safety systems.

## 2.     What Are 'Intelligent Systems'[1]

The debate about what constitutes an 'Intelligent System', an 'Expert System', a 'Knowledge Based System' or 'Artificial Intelligence' is one that has raged since these terms were invented.  For the purposes of this paper, we will use these terms interchangeably and use the following definition:

> An Intelligent System (or Expert System, Knowledge Based System, AI system) is one that incorporates a significant amount of information or knowledge that has been gleaned from the domain in which it is applied.

This definition will, no doubt, raise the hackles of some of those operating in the AI field, but will be good enough for the purposes of the paper.

The important aspect of this definition is that a system uses previous knowledge of its domain.  The knowledge may be gleaned by interviews with domain experts and encoded in rules (as in Rule Based Systems), it may be deduced from large examples (as in Rule Induction Systems or Neural Networks) or it may be gained directly from the system's own environment.  However, to be an Intelligent System it must use this knowledge in its operation.  Typically, the knowledge will be used to guide the system to an acceptable solution, discarding less promising routes on the way.

At present, the problem of assessing these systems from a safety viewpoint is not insurmountable.  Techniques for verification and validation are available, albeit that they require transparency of the system under evaluation; many of these techniques will already be familiar to anyone involved in the assessment of software systems.  However, the assurance gained from these V & V tools diminishes in parallel with decreasing system transparency, and it is this problem that we wish to address in this paper.

## 3.     Basic Types of Intelligent Systems

### 3.1.     Introduction[2]

Developments in AI over the last thirty to forty years have yielded a huge range of systems, making it difficult to summarise their relevance and impact on safety

---

[1] The term has changed subtly from "Artificial Intelligence" to "Intelligent System".  This reflects a growing favour for the use of the term "Intelligent System" in place of "Artificial Intelligence" as being more appropriate to current concepts.  In addition it is better suited to the tenor of this paper, as will become clearer later on.

[2] This overview is not intended as a detailed introduction to AI, and is certainly open to criticism of over-generalisation from AI experts.  It is hoped however, that it will give enough of a picture to place the remainder of this paper into context.

engineering in just a few paragraphs. The different methods and techniques can be classified along a number of dimensions, but the transparency of the way in which a system operates is of crucial importance to safety analysis. Consequently, we have chosen in this paper to present brief descriptions of a range of some AI techniques arranged in a roughly decreasing order of transparency.

This transparency of operation is associated to the degree of predictability in these various types of system. In between the extremes of total predictability and total unpredictability are the many systems that may be theoretically predictable, but are effectively unpredictable because of the complexity of either their structures, their inputs or both. However, in this context the more generalised concept of transparency is preferred as being more suitable to the discursive nature of this paper.

## 3.2.    Rule Based Systems

Rule Based Systems are perhaps the best understood and the most widely implemented of AI techniques. They operate by the selection and 'firing' of a series of IF... THEN... rules, arranged in a rule base. Typically, a system will look at the evidence it has available to it and will use the rules to infer other true facts about the situation. (Alternatively, a system may be told what it should prove, find those rules in its rule base that can reach such a conclusion, and then seek to prove the 'IF' part of those rules, iteratively seeking more information until the original premise is supported or discarded). In most cases, the rules are written in a form of English (or French or Italian or .......) and can be readily understood by a domain expert. For example, the rule:

> IF          Flow Speed is greater than 50 litres/min
> *and* Temperature is greater than 48°C
> *and* System Status is Active
>
> THEN        Recommend 'Reduce Flow Speed'
> *and* Set System Alarm = 'Amber'

could be easily understood by most industrial engineers.

Rules are therefore generally open to analysis from a safety viewpoint, although inconsistencies in the rule base may not be apparent, and large rule bases may render meaningful analysis difficult, if not impossible. Rule bases may also be incomplete and fail to cope with some combinations of inputs, and this may not be apparent from a simple inspection of the rules. The use of weighting factors in rules (a common technique) further undermines the efficacy of currently available V & V techniques.

## 3.3.    Rule Induction Systems

Rules may be developed by experts in the domain, or may be generated by Rule Induction Systems. These systems take a set of data and seek out underlying rules that characterise the data. These rules may then be used by a rule based system acting on real data. The rules from induction systems vary in their intelligibility, but are also generally understandable by a domain expert.

## 3.4.    Constraint Propagation

Constraint Propagation is a technique whereby all the constraints in a set of related objects or events are specified, and the system then tries to come up with a solution to a problem that satisfies all these constraints. An example may be a system for creating school timetables which may have constraints such as:

> Classroom A can hold only 30 students.
> It is not possible for two classes to be held in a classroom at the same time
> Group B has 46 students
> Safety Engineering must be taught before lunch

The system will use these constraints to find a compliant solution (or number of solutions).

The constraints can generally be understood by a domain expert, but similar problems to those experienced with Rule Based Systems will arise if the constraint base is of significant size. In addition, Constraint Propagation Systems may fail to reach a valid conclusion, thus resulting in some form of system 'freeze'.

## 3.5.    Case Based Reasoning

Case Based Reasoning is a relatively new technique, which operates by using previously solved problems to solve new ones. Typically, when the system is faced with a problem, it will search for a previous Case (Problem/Solution pair) with a similar or identical problem. If an identical match can be found, then the previous solution can be used. If only a close match can be found, then the previous solution may be adapted to solve the new problem. If this is successful, then the new problem/solution pair may be added to the Case Base for later use.

From a safety point of view, this system is superficially attractive, as only previously successful (and presumably acceptable) solutions are presented. It is possible, however, that some part of the new problem may not be captured in the Case Base, and that the two apparently identical problems may actually be different. Thus the old solution may fail in practice when applied to the new situation.

Depending on how the problem is captured and described however, the various individual cases should be suitable for review by experts to assess their safety.

## 3.6.    Genetic Algorithms

Genetic Algorithms represent a class of solutions based on an analogy with development by natural selection. In a Genetic Algorithm system, a large number of potential solutions to a problem are generated and tested against the problem to hand. The less successful solutions are discarded and the remainder are 'interbred' to generate a new set of possible solutions. This new generation of solutions is added to the pool of solutions and the cycle repeated, this cycle being continued until a suitable successful solution is found.

Genetic Algorithms are typically not well suited to analysis and are often applied in domains where the effect of changing a particular parameter of a solution is not easy to predict.

## 3.7.    Neural Networks

Neural Networks are constructed using a large number of simple processing units massively interconnected, such that the output of any unit will typically feed the input of many others. Some inputs are then 'connected' to some measurements or characteristics of the problem domain, and some outputs are designated to indicate some analysis of that domain.

The system is then exposed to a large number of known input/output sets, and by changing the weight of connection between the various processing elements using suitable mathematical techniques (or 'magic' to the non technical), the system is trained so that its outputs correctly match those in the training set.

From a safety viewpoint Neural Networks have some attractive features: they are robust, capable of generalising in the face of missing or noisy inputs and capable of reacting appropriately to previously unseen situations. Their operation is however, completely opaque and lost in the massive interconnection of many non-linear elements. The logic behind their decisions is not therefore accessible to an analyst.

## 3.8.    Summary

'So what's new?', we hear conventional programmers cry. 'We've been encoding domain knowledge in IF/THEN structures for years!' Nothing, of course, is new; the overlap between Intelligent Systems and conventional systems has merely become more blurred. Indeed, in any Intelligent System, a significant core of the program will be nothing more than standard code, responsible for the usual data manipulation found in any computer system.

This brings us to some general lessons which will be of value in the assessment of Intelligent Systems with regard to safety:

> a.    We cannot forget all the baggage that comes with safety assessment of conventional software systems when we encounter an Intelligent System. Indeed, it is possible that an AI developer will be less capable of (and perhaps less interested in) putting together the conventional elements of the parent system than would a conventional programmer. This may mean that we need to be more stringent in our assessment of safety systems at system level, and of the conventional elements in particular.

b.        We also need to explore the knowledge encoded in the system. This may be more or less open to analysis than in a conventional programme. It may also contain bugs of a new kind. A rule based system may have a clearly defined rule base that is open to inspection; however, it may be that although every rule in the system is individually correct, when used in combination errors emerge. For example, some critical rules may not be reachable and will never be 'called' by other rules; or the rule base may be inconsistent; or the rules may allow the system to conclude that a fact is both true <u>and</u> not true at the same time. Other problems may include redundancy and circularity.

c.        The knowledge encapsulated in the Intelligent System representation may not be open to inspection. For example, in a neural network, the 'knowledge' is encoded in the relative weights of the mass of interconnections between the elements. Although these weights are not hidden from the analyst, they are individually meaningless. This renders exhaustive analysis futile, and means that a new approach to the assessment of system safety is required.

# 4.        Learning systems

## 4.1.        Introduction

There is a further significant subset of Intelligent Systems that pose additional difficulties for safety engineers. These are systems that learn over time. We may apply all the V + V techniques that we wish on the system that we place into service, but if the system starts to change as soon as it is being used, then confidence in the safety of the system will decline (it must be recognised that change control is a fundamental requirement of a safety management system).

The learning may occur in a number of forms, some more hazardous than others. In a rule based system for instance, we may add new rules to overcome perceived shortcomings in the system. These could be checked against the remaining rules (effectively a check of the entire rule base, or some significant subset of it) before they are implemented. Whilst this allows us to retain some control of the safety of the system, it would reduce the scope or speed at which the system could change. A complete rule base analysis may be required after changing only one rule, so system updates may only be cost effective when a group of new rules can be 'batched up' for checking before implementation of a new rule base. This negates the speed at which the system can evolve, thus diluting one of the key attractions of learning systems.

More serious problems are presented by systems that learn in real time, without reference to the system designers. In such systems we cannot be sure of the state of a system (and hence its safety) once it has been in operation for any length of time.

The only way we can tell how such a system is operating is to observe its reaction to inputs, which may give us some confidence, albeit this confidence may be misplaced.

## 4.2.   Safety Training of Intelligent Systems

By their nature safety critical systems generally operate within safe, central conditions and it will be these conditions that the system is using to 'train' itself. This training may feasibly be at the expense of the system's reaction to unsafe inputs, and _these_ reactions are exactly those that are of interest to us as safety engineers. Some mitigation of this effect may be gained by sensible design of the learning system. Illustrative examples of some approaches are given below.

A system's learning process could be designed as follows:

> Next time you do something, allow the system to move a little further outside the bounds you have previously used. If this results in a safe action, then use these new bounds in your future operations. If it is dangerous, do not use the new bounds.

We could call this the Toddler model. A toddler is generally cautious but expands its operations until it meets a perceived danger (a hot fire, a step to fall down, running too fast). Minor scrapes will teach a toddler how far he/she can go, but step changes from safe to unsafe, or hazards outside the toddler's knowledge or perception (eg live electrical sockets), could result in disaster.

Another possible learning process could be:

> Next time you do something, vary the way in which you do it (within the safety boundary) and monitor how safe it was. If it was less safe than you expected, don't do it again. If not, then change the bounds of your operation to allow it.

This could be called the Coward's model. If a coward tries something new and it frightens him, then he will not try it again.

The key differences between these two approaches are two-fold:

> a.      The Toddler's model is not initially aware of its safe boundaries, and is only successful if the system can withstand occasional, limited excursions into unsafe areas, perhaps protected by a parallel system (parent) that shuts down operations if the system exceeds its safety envelope. It does, however, encourage a complete exploration of the operating space, which (if disaster does not befall the toddler along the way) will lead to an efficient, capable system.

b.    The Coward's model by contrast is acutely aware of its safety boundaries, and veers away from danger, making the system safer. However, it also constrains the scope of operations of the system and may miss the most capable methods of carrying out its task. This model also relies on some variable measure of the level of hazard risk that it perceives.

In general, the best solution lies somewhere in between these two extremes, allowing the 'Toddler' to push the boundaries of the envelope and the 'Coward' to start turning up the volume on the alarm bells as danger approaches.

# 5.    Using Intelligence in Safety Related Systems

## 5.1.    Introduction

In order to discuss the validity of using intelligence in safety critical systems, a brief examination of how such intelligence may be used is necessary.

## 5.2.    Intelligence as a Functional Element of a System

Whether used as serial components (where the system relies entirely on the element) or in parallel with other elements (be they non-intelligent, intelligent or human), such intelligent systems would perform a function that is vital to the successful operation of the system. This operation may be on-line or standby, or in some form of shared operation (eg voting systems).

## 5.3.    Intelligence as a Supervisory or Monitoring Element of a System

Intelligent systems can be used as monitors or supervisors of other systems (or of other elements of the same system). In such cases they would not interfere directly with the functions of the system, but would record and analyse these processes, presenting this in quantitative and qualitative terms, if necessary in near real time. Such a usage appears to presuppose a human involvement at some point in the system, albeit that this may be at some remote site, although this need not necessarily be so. The concept of a safety manager that can override a non-intelligent or lower-intelligence operating system has already been envisaged, and indeed is probably already incorporated into a current system. The function of such a system would be to report transgressions of the system outside safety boundaries, and if necessary veto such operations.

Of course, the roles of intelligent and non-intelligent systems in this manner can be reversed. It is entirely feasible that the functional element could be intelligent, whilst the safety monitor is non-intelligent. Either way, the system can be designed such that only one of these systems need be designed to high-integrity standards; the designer, of course, has the choice of which one.

## 5.4. Intelligence in Support of System Design and Analysis

Aside from the use of AI as a system component, there has been significant progress in its use during design and evaluation of safety related systems. Intelligent systems provide the safety engineer with valuable knowledge-based tools; the use of expert systems for verification and validation, or for use in FMEA studies (such as Ulster University's Fire Safety Evaluation Scheme [Donegan/Taylor 93] and Aberystwyth University's FLAME [Pugh et al 95]) are examples. Whilst acknowledging that this is a very important application of synthetic intelligence systems, we do not intend to address the use of AI in design and analysis tools in any detail in this paper as our key theme is the use of intelligence as a system component.

However, many systems have become so complex and sensitive to trivial input changes that complete analysis becomes a virtual impossibility. Although AI can support this process, there is also a move towards analysis whereby a system can be treated as a black box where only the interface performance is assessed, that assessment being achieved by comparison with benchmarks provided by an 'acceptable' system. The implications of this need to be watched carefully.

## 5.5. Parallel Use Versus Serial Use

Should system designers actually use intelligence as an unsupported serial function in a safety-related system if the safety case for such usage is difficult or impossible to make? This argument is a key point, and it applies equally to the use of human as to synthetic intelligence. Traditional AR&M theory shows that parallel duplication of functions results in a marked improvement in overall function availability, particularly where common-mode failure points can be eliminated. Such redundancy is often used when it would not be cost-effective to improve the reliability of the individual function element.

AI can be used in this fashion in various ways: in parallel with another intelligent system; in parallel with a non-intelligent system; in parallel with a human. The primary functional element could be either intelligent or non-intelligent, and could still be subject to full hazard analysis processes. The addition of parallel systems should not degrade the safety case but would serve to enhance it, thus reducing the overall system risk. Such synergy must be beneficial and should further reduce the risk through the system life.

This concept that demonstrable benefit (in terms of reduction of risk) can accrue is another key argument. If we progress towards 'black box' analysis against given acceptable benchmarks (as described above), then empirical and qualitative analysis can be used rather than exhaustive quantitative analysis. This would be invaluable in cases where exhaustive analysis would be impractical.

# 6. Comparison of Synthetic and Human Intelligence

## 6.1. Introduction

One can argue that the term Artificial Intelligence is an oxymoron, in that if something is truly intelligent, then its origins are irrelevant (cogito ergo sum?). In this respect artificial (or better, synthetic) and human intelligence should be capable of being addressed together. Synthetic intelligence such as this can be defined as a system that expresses all the aspects of intelligence that we see in humans. This would, presumably, require the synthetic intelligence to pass some form of criterion akin to the Turing test [Turing 50].

In consequence, as synthetic intelligences approach human intelligence in capability (and in many cases well before this), they will be seen to display certain common attributes, such as: an ability to learn (obviously), an ability to cope with unforeseen circumstances, inductive reasoning and intuitive reasoning. To some extent, we are already witnessing these attributes in synthetic intelligent systems. Not all is plain sailing, however. As we have seen from their make-up, intelligent systems will also have certain detrimental aspects in common with their human counterparts, such as:

> lack of transparency as to how and why they react to given conditions,
>
> change of behaviour over time,
>
> unknown reactions to extreme inputs.

However, humans also suffer from characteristics that have proved to be extremely difficult to eliminate (whether by education or training), but which could probably be handled better using synthetic intelligence, such as:

> degraded performance due to fatigue, boredom or intoxicants,
>
> fixations on certain hypotheses (mind sets),
>
> emotive reactions,
>
> quixotic or illogical behaviour,
>
> unreliable memory.

## 6.2. Human Reliability

In particular, synthetic intelligence is likely to prove (in overall terms) a significantly more reliable performer than its human counterpart - certainly no worse.  To assist in the justification of this claim it is worth reviewing the table of 'Human Error Probabilities' from Interim Defence Standard 00-56 Issue 1 [Def Stan 00-56/1] (regrettably omitted from Issue 2) an extract of which is reproduced in Table 1.  Although produced originally for a specific use, this data should equally apply to any situation where the human consciously faces death or serious injury (either personal or to others) when performing as part of a system.

| Nature of Task | Failure Probability[3] |
|---|---|
| General omission error, when there is no warning alarm or display | $10^{-2}$ |
| Errors of omission when the actions are embedded in a well-rehearsed procedure | $3 \times 10^{-3}$ |
| General error of commission | $3 \times 10^{-3}$ |
| Simple arithmetic errors with self checking | $3 \times 10^{-2}$ |
| General error of supervision | $10^{-1}$ |
| Handover/changeover error | $10^{-1}$ |
| General decision error rate for high stress levels | 0.2-0.3 |
| Failure to act correctly in reasonable time after the onset of a high-stress condition | 0.3-1.0 |

Table 1

Thus, whilst synthetic intelligent systems acquire more of the recognised human intelligence qualities, they are also likely to exhibit significant advantages over their human counterparts.

As intelligent systems further develop, the distinction between synthetic and human intelligence may become even harder to define.  A certain AI school (to which the authors would subscribe) will tell you that there is no theoretical reason why a computer should not exhibit all the characteristics we associate with human intelligence, such as emotion, humour, sympathy, flexibility, perception, pleasure, the willingness to cause harm etc.  This premise is fiercely contested by others however, who claim that some of these aspects of human intelligence are unique to humans [Descartes 1637], [Searle 80], [Penrose 89].

This paper is not the place to resolve these arguments, which have in any case been raging for centuries!  We can however speculate on the way in which intelligent

---

[3] These figures assume a competent adult in good health, and not under the influence of drink, drugs, tiredness or boredom.

systems may develop. As greater advances are made, processing becomes smaller and cheaper, and it is possible that machine intelligence will converge with that of humans. What might this mean for safety engineering? Our contention is that this could mean that the way we treat machines in safety related systems will have to converge with the way we treat humans. Just as we do not currently insist on disassembling a human brain to see if it is suitable for controlling a potentially hazardous system, so we may not insist on delving into the depths of a computer.

# 7. Some Problems of Intelligence in Safety Systems

## 7.1. Introduction

We have looked (briefly) at the fundamental problems posed by the need to achieve safety assurance for intelligent systems. There are further problems that may arise, however, where even simple approaches and rules can lead to complex safety situations. A couple of examples from literature and life are pertinent:

Isaac Azimov, as one of the proponents of intelligent robots, claimed in many of his novels that synthetic intelligence would approach, if not equal or exceed, that of human beings. He did however instil in his robots the three 'Laws of Robotics[4]' [Azimov 92 et alia]:

> 1. A robot may not injure a human being or, through inaction, allow a human being to come to harm.
>
> 2. A robot must obey the orders given it by human beings except where such orders would conflict with the First Law
>
> 3. A robot must protect its own existence as long as such protection does not conflict with the First or Second Law.

These rules appear quite reasonable and self-evident, but they are in fact simplistic. As Mr Azimov himself acknowledges, application of the laws of robotics leads the robot into a complex argument of degree and probability. Can a robot policeman harm a potential criminal armed with a lethal weapon in order to protect a greater number of innocent people? Can a robot surgeon operate on (ie harm) a human being in order to save that human's life? Faced with such dilemmas, the intelligent system must evidently perform both qualitative and quantitative analysis of the

---

[4] As these laws stand, they deny Mr Azimov's robots the ability to possess the full equivalent of human intelligence, in that no human intelligence is inextricably bound by similar laws, irrespective of strict social conditioning, education or training. All human beings, however conditioned, possess the innate ability to behave irrationally and thus harm others or themselves, and can exercise this ability virtually at will. Now, from a safety point of view, it is arguable that these last aspects of human intelligence are in fact manifestations of failure modes in the human psyche; this would of course be a simplistic view, just as Azimov's three laws are themselves simplistic.

situation in order that the optimum solution is performed, noting that the 'do nothing' solution is very often equally unacceptable.

As a further example, consider the case of the introduction into a hospital of an intelligent system to analyse (say) kidney function. In its domain, the machine may be as proficient as a qualified and experienced consultant, but this may still lead to two sorts of litigation. A practitioner may be criticised for using the machine rather than his own expert judgement, whereas a junior doctor who used his own judgement could, perhaps, be sued for negligence for not using the best tools available to him. Who is to say where the boundary between these two cases should be drawn?

## 7.2. Using Human Intelligence in Safety Systems

It would thus appear that in assessing the safety of our learning synthetic intelligent system, we come across the problems that are faced every day in assessing the hazards involved in letting humans use systems. How then are we to assess whether the machine is fit to perform the safety function previously allocated to a human? The logical progression of the previous argument is to ask the question: 'If an intelligent system becomes so advanced that it may be indistinguishable from human intelligence, then should we not treat such systems in the same manner that we treat human systems?'

This however leads us to an important problem, and one which can be argued is a serious flaw in our current safety philosophy. This is the assumption that our current treatment of human intelligence within a safety critical system is satisfactory. Consider this dilemma: an intelligent system is developed to drive a car. This system possesses a capability that demonstrably exceeds its human counterpart in capability (reaction times, response accuracy, interpretation of conditions, planning etc), and is developed to a level of reliability that is also better than its human equivalent (such systems may well be currently under consideration, or even development [Barber/Smith 96]). Thus we have our intelligent car-driving system (robot, if you will), which is not only demonstrably more capable than a human, but will probably offer significant added value as well. Are we then going to allow that system to drive our cars with no further testing other than the Department of Transport basic driving test? Of course not.

On the contrary, we would probably insist on some of the most exhaustive testing possible, and with the sort of rigour that probably could not be achieved with our current tools and methods - at least, not within an acceptable timescale and budget. The safety flaw referred to above is of course that we currently allow humans to perform such safety functions with only minimal checks on their capability, training and ongoing performance. Defence Standard 00-56 is worth consulting again: 'Due regard shall be taken of human fallibility wherever a safety feature is implemented by a human being. Human factors, instructions and training shall be considered when apportioning safety integrity ........'. Is the absence of this consideration perhaps why we kill 10 people a day on British roads, and seriously injure many more?  No safety inspectorate would even contemplate the introduction of a novel system that claimed a fraction of such statistics;  the public expectation and utility of private road transport has nevertheless permitted this situation to come about, and we are for the time being stuck with it.

This paper was not written with the purpose of denigrating road transport; the example was chosen to demonstrate a key point about the assessment of intelligent systems, which is indeed the nub of the paper.  This is:  is it strictly necessary to demonstrate the reliability of an intelligent system in exhaustive detail, when it can be demonstrated by empirical or experiential analysis that such a system would be a significant improvement on an existing alternative that is already operating to an acceptable safety level?

## 7.3.    Replacing Human Intelligence in Safety Systems

We must ask ourselves the question: is it a logical or correct assumption that the designer should attempt to replace the human in a system?

On the one hand, it is self-evident that the poor reliability of the human, alluded to above, sometimes simply does not match the expectations of the designer, in particular in strictly deterministic systems where the output must be both reliable and repeatable.  For example, it has proved beneficial to remove humans from systems where speed and reliability of response is required over long periods of time, from systems containing tasks that the human finds repetitive and tedious and from situations which would place the human at unacceptable risk.  In such cases, replacement of the human is both desirable and beneficial, although the replacing system itself may not be of high intelligence.

However, although this argument can be, and has been, used in systems with low intelligence, can it also be extended to those containing higher orders of intelligence?  Clearly, if a safety case has been made for a system involving a human as a functional element, <u>and</u> it can be shown that a synthetic intelligence can perform the required functions to a greater degree of reliability and safety, then it would appear to be reasonable to suppose that the human could be replaced.

Notwithstanding the logic of this argument, public perception has in the past often overruled the designer in favour of the human.  For example, automatic, computer-controlled train systems have been given a human 'minder' because passengers were uncomfortable about riding in unmanned trains.  Attempts at introducing automated medical diagnostic and banking systems have similarly fallen foul of public opinion.  Instead, the public seems to be happier with some form of partnership arrangement, irrespective of whether the human or the synthetic intelligence takes precedence.  Such partnerships could well prove to be invaluable as intermediate stages, during which the synthetic intelligence could prove itself to be equal (or probably superior in some aspects) to its human partner, thus gaining general, and specific, approval for use as a sole functional element.

## 7.4.    Controlling Change

The need to monitor the control of change in a system presents a considerable challenge to the introduction of intelligence into safety systems.  Systems that learn, and subsequently change themselves as a result of that learning, can negate the basis on which the original safety case was built.  We would be unwise to assume that an intelligent system will only improve through learning; after all, this has not been our experience with human intelligence.  Change can also result from degradation - which we have experienced with non-intelligent systems that can drift out of tolerance, but which may be due to sensory deterioration (for example, eyesight and hearing in humans) or from processing deterioration (from processor or memory overload to narcotic influence).

Clearly, there must be some means built into either the intelligent system, or into the parent system that monitors and takes account of change.  Due allowance for it must be made in the system's operation, even to the extent of disabling the system if the change takes the system outside safety boundaries.  In extreme cases, it may be necessary to prevent change from occurring until there has been a complete review of the system's safety case, although this could negate the benefits of intelligent learning systems.  Whatever the solution, this is a subject which will surely keep philosophers, designers and regulation writers employed for some time yet!

# 8.    Conclusion

Intelligent systems are with us, and barring some form of Luddite revolution, they will in due course approach, if not exceed, the capability of their human counterparts in many areas.  Although the safety engineering community has started addressing the use and assessment of intelligence within safety related systems, there is still a shortfall in our understanding of the implications of using higher orders of intelligence, in particular in the regulation of this use, and in the guidance given to the designer and safety manager.

It would be simple to proscribe the use of certain levels of synthetic intelligence in safety engineering, but such over-conservatism would undoubtedly deny the engineer and designer the use of enormous potential capability.  However, a cautious approach is an inherent requirement in safety engineering.  We must therefore avoid the cavalier use of intelligence for its own sake and must justify it on the basis of enhanced overall system performance, whether its use replaces or supplements humans within the system.

As techniques for the evaluation of safety systems develop in parallel with the systems themselves, we must constantly review the requirements needed to provide an auditable safety case.  The cost-effectiveness of such techniques must also be examined, to ensure that the gain in system utility provided by AI warrants the increased expenditure in consequential analysis and risk reduction.  It is the contention of this paper that the time to review such considerations is imminent, and that we in the Safety Critical Systems community should be actively engaged in formulating a policy on how designers are to address the problems that will be presented by intelligent systems.

As a closing note, we would like to offer the following extract from a newspaper report of an airmiss investigation in the year 2097:

> 'The pilot, an AttoPilot 14x, number 31415927 (known to friends as Pie7), had followed a normal career pattern up to this point. Following manufacture at the IBMITO plant on the India/China border, she was trained on a Twin Engined PseudoProp in cyberspace area 124d3q (a common training area for air and space pilots). She passed her pilot's exams following 3000 hours on the New York/Los Angeles international route, and was certified to fly this 2700 seat Aerospat/BritAir/Daimler/ Naaihto/Virgin 7237 four years ago. She has been piloting small passenger aircraft such as this continuously since this time, much of it in the crowded skies above Eurafrique. The investigation confirmed that she had been loaded with all relevant practices, and that the weather was under control at the time of the accident.

> The investigation board, which consisted of 11 personnel, 5 remote processes and an experienced safety engineer (a Type 4), concluded that the incident was caused by Pie7's excessive use of a hallucinocarcinogenic software agent, widely abused by airline software, but generally considered benign. The pilot, who has been grounded since the incident, was ordered to undergo pseudopsychological counselling, and will have to undertake further training and requalification before she is allowed to pilot another passenger aircraft.'

# 9.    References


[Azimov 92 et alia]    Azimov, Isaac (1992) "The Positronic Man", Victor Gollancz Ltd, Preface.

[Barber/Smith 96]    P A Barber & D P Smith "Effects of Technology on the Safety of Automotive Transport", Proceedings of the Fourth Safety-Critical Systems Symposium, Leeds 1996, pp 266-284.

[Def Stan 00-56/1]    Interim Defence Standard 00-56 / Issue 1, 5 April 1991, "Hazard Analysis and Safety Classification of the Computer System Elements of Defence Equipment", page 27 table 9.

[Descartes 1637]    R Descartes (1637), "Discourse on Method".

[Donegan/Taylor 93]    H A Donegan & I R Taylor "The Potential of Expert Systems in Fire Safety Evaluation", Journal of Applied Fire Science Volume 3, Number 4 - 1993-94, pp 315-333.

[Penrose 89]    Penrose, Roger (1989), "The Emperor's New Mind", Oxford University Press.

[Pugh et al 95]    D R Pugh, C J Price and N Snooke "Practical Applications of Multiple Models - the Need for Simplicity and Reusability", in "Applications and Innovations in Expert Systems", the Proceedings of Expert Systems 95, the Fifteenth Annual Conference of the British Computer Society Specialist Group on Expert Systems, Cambridge, December 1995.

[Searle 80]    J Searle (1980), "Minds, Brains and Programs", Behavioural and Brain Sciences, Volume 3, Cambridge University Press.

[Turing 50]    Turing, Alan (1950) "Computing Machinery and Intelligence", Mind 59, pp 434-460.

# How to Improve
# Safety Critical Systems Standards

Norman Fenton
Centre for Software Reliability
City University
London

## Abstract

An effective standard for safety critical software systems is one that should help both developers and assessors of such systems. For developers it should be clear what is required in order to conform to the standard, while for assessors it should be possible to determine objectively compliance to the standard. The existing set of standards do not pass this basic quality test. We provide a framework for improving such standards. We view a standard as a collection of requirements. For each requirement we first identify the process, product, or resource that is the primary focus. Next we consider the clarity of the requirement in respect of the ease with which it is possible to assess conformance to it. We describe guidelines for interpreting requirements to make this task more objective. The method is applied extensively to the IEC 1508 Safety Critical Standard.

## 1 Introduction and Background

Between 1990 and 1994 researchers at CSR City University were involved in a collaborative project (SMARTIE) whose primary objective was to propose an objective method for assessing the efficacy of software engineering standards [Pfleeger et al 1994]. The method was based on the simple principle that a software standard is effective if, when used properly, it improves the quality of the resulting software products cost-effectively. We considered evidence from the literature and also conducted a small number of empirical studies of specific company standards. We found no evidence that any of the existing standards are effective according to our criteria. This will come as no surprise to anybody who has sought quantitative evidence about the effectiveness of *any* software engineering method or tool. However, what concerned us more was that, in general, software engineering standards are written in such a way that we could never determine whether they were effective or not.

There was certainly no shortage of standards to review. We came across over 250 standards (from various international and national bodies) that we considered to fall within the remit of software engineering. The common feature of all of them was that they define some aspect of perceived 'best practice' relevant for developing or assuring high quality software systems or systems with software components. Unfortunately, there is no consensus about what constitutes best practice, and it follows that there is no consensus as to how to distinguish those best practice techniques that should always be applied. Thus, for standards of similar names and objectives we came across very different models of software quality and the software development process. This was especially true of the safety critical software standards; of which IEC SC65A [IEC 1992] and DEF-STAN 00-55 [MOD 1991] were two significant examples.

We discovered the following general problems in the standards we reviewed:

1) *Heavy over-emphasis on process rather than product.* Traditional engineering standards assure product quality by specifying properties that the product itself must satisfy. This includes the specification of extensive product testing. Software standards almost entirely neglect the product and concentrate on the development process. Unfortunately, there is no guarantee that a 'good' process will lead to a good product, and there is no consensus that the processes mandated in many of the standards are even 'good'.

2) *Imprecise requirements*: The traditional notion of a standard is of a set of mandatory requirements. Such requirements must be sufficiently precise in definition so that conformance can be determined objectively by appropriate tests. Where no such precision is possible, and hence where mandatory enforcement is impossible, standards bodies traditionally defined documents as 'codes of practice' or 'guidelines'. In this respect software engineering is subjected to a proliferation of 'standards' which are at best guidelines that could never be mandated realistically Another basic property of any good standard is that an independent assessor should be able to determine if the standard has been applied or not. We found that a vast majority of requirements are presented in such a way that it is impossible to determine conformance in any objective sense. Thus in general it would be impossible to determine whether or not the standard has been applied. This makes a mockery of many of the assumed benefits of standardisation.

3) *Non-consensus recommendations*: Many of the standards prescribe, recommend, or mandate the use of various technologies which have not themselves been validated objectively. The standards may therefore be mandating methods which are not effective for achieving the aim of high quality systems.

4) *Standards too big.* Most standards attempt to address the complete system development life-cycle. This results in extremely large documents containing sets of un-related requirements, of which many will be irrelevant in a given application. Such standards are almost impossible to apply, and generally stay 'left on the shelf'.

In this paper we propose a framework for improving standards. The approach (which is based much on the SMARTIE philosophy) is applicable to *any* software standards, but is especially pertinent to the safety critical ones. The latter can be viewed as simply the most demanding of the software standards; if you remove the safety integrity requirements material from such standards then they can be applied to any software system with high quality requirements. Our framework for interpreting standards is to view a standard as a collection of requirements that developers have to comply with and to which assessors have be able to determine conformance. In Section 2 we discuss the notion of clarity and objectivity in these respects. Our objective is to provide recommendations on how to rationalise and refine standards in such a way that we move toward the scenario where at least the obligations for the assessor are clear and objective. In Section 3 we explain how to classify requirements according to whether they focus primarily on one of three categories: process, product, or resource. Using this classification, we show how the safety critical standards concentrate on process and resource requirements at the expense of clear product requirements. We explain how to shift the focus toward the product requirements. In Section 4 we explain how requirements could be interpreted in such a way that there is greater objectivity, especially for the assessor.

Our emphasis is on how we can interpret and use standards *despite* their current weaknesses. We do not question the importance of standards to safety critical systems development. However, clearly some standards are better than others and some requirements are more important than others, even though *a priori* we do not know which. Thus in Section 5 we discuss the need for assessing the effectiveness of standards, and describe the basic principles behind a measurement-based procedure.

Throughout the paper we concentrate on the recently issued, and highly significant IEC1508 [IEC 1995] (of which Parts 1 and 3 are relevant) as an example of applying our method. This standard is the updated version of IEC SC65A.

## 2 Clarity of Requirements in Standards

A standard is a collection of individual requirements. Our main concern is to consider the clarity of each mandatory requirement in the following two keys respects:

1. *The developer's obligations for compliance*: is it clear what is required in order to conform to the requirement? If not then the standard cannot be used in a consistent and repeatable way.
2. *An assessor's obligation for determining conformance*: is it possible to determine conformance to a requirement reasonably objectively? If not then we may not be able to trust the assessor's results.

Generally, obligation (2) will follow from (1). For example, in IEC 1508[1], Part 1, there are a number of requirements concerning the Safety Plan. Of these 6.2.2e asserts: that the Safety Plan shall include a 'description of the safety lifecycle

---

[1] unless otherwise stated all examples are taken from IEC 1508 Part 3

phases to be applied and the dependence between them'. The developer knows that certain specific information must appear in the document. The assessor only has to check that this information is there.

Conversely, however, it is not necessarily true that (1) will follow from (2). For example, for the software safety lifecycle we have:

*Requirement 7.1.6*: "Each phase shall be terminated by a verification report"

Obligation (2) is clear. The assessor has, strictly speaking, only to check the existence of a specific report for each specified phase. However, the developer's obligations for the requirement is unclear; a subsequent requirement (in the software verification section) sheds little light on what constitutes an acceptable verification report:

*Requirement 7.9.2.4:* "A Software Verification report shall contain the evidence to show that the phase being verified has, in all respects, been satisfactorily completed."

Unfortunately, in a key standard like IEC 1508 most requirements are unclear in both respects. For example, requirement 7.4.6.1a asserts that:

"The source code shall be shall be readable, understandable and testable"

It is unclear what is expected of developers, while an assessor could only give a purely subjective view about conformance.

In traditional engineering standards it is widely accepted that the necessary clarity for both obligations (1) and (2) have to be achieved for all requirements [Fenton et al 1993]. Partly because of the immaturity of the discipline, software engineering standards do not have this clarity. Our objective here is to provide recommendations on how to rationalise and refine standards in such a way that we move toward the scenario where at least the obligations for the assessor are clear and objective.

# 3 Classifying requirements in standards

## 3.1 Processes, Products, and Resources

Our approach to interpreting standards begins by classifying individual requirements according to whether they focus primarily on *processes*, *products*, or *resources*:

A **Process** is any specific activity, set of activities, or time period within the manufacturing or development project. Examples of process requirements are:

*7.1.4*: "Quality and safety assurance procedures shall run in parallel with lifecycle activities" (process here is *Quality Assurance*)

*7.4.8.7* "Test cases and their results shall be recorded, which may be in machine readable form for subsequent analysis" (process here is *Testing*)

*7.9.2.12* "The source code shall be verified by static methods to ensure conformance to the Software Module Design Specification, the Coding Manual, and the requirements of the Safety Plan (process here is *static analysis*)

A **Product** is any new artefact, deliverable or document arising out of a process. Examples of product requirements are:

*7.2.2.5a* "The Software Requirements Specification shall be expressed and structured in such a way that it is as clear, unequivocal, verifiable, testable, maintainable and feasible as far as possible commensurate with the safety integrity level" (product here is *Requirements Specification document*)

*7.4.6.1b* "The source code shall satisfy the Software Module Design Specification" (product here is source code)

*7.4.8.5* "The Software Integration Test Report shall be in a form such that it is auditable" (product here is the *Software Integration Test Report*)

A **Resource** is any item forming, or providing input to, a process. Examples include a person, a compiler, and a software test tool. Examples of resource requirements are:

*Part 1, 5.2.1* "All persons involved in any life-cycle activity, including management activities, shall have the appropriate training, technical knowledge, experience and qualifications relevant to the specific duties they have to perform" (resource here is *people*)

*7.4.4.3a* "The programming language selected shall have a translator/compiler which has either a 'Certificate of Validation' to a recognised National/International standard or an assessment report which details its fitness for purpose" (resource here is the *programming language compiler*)

*7.4.4.3b* "The programming language selected shall be completely and unambiguously defined or restricted to unambiguously designed features" (resource here is the *programming language*)

*7.7.2.7* "Equipment used for software validation shall be calibrated appropriately and any tools used, hardware or software, shall be shown to be suitable for purpose" (resources here are *tools*)

Ideally, it should be absolutely clear for each requirement which process, product, or resource is being referred to and which property or attribute of that process, product, or resource is being specified. The example requirements above are reasonably satisfactory in this respect (even though they do not all have the desired clarity discussed in Section 2). However, in many requirements, it is necessary to 'tease out' this information. Consider the following examples,

*7.4.5.3* "The software should be produced to achieve modularity, testability and maintainability"

Although this refers explicitly to the software production *process*, this requirement really only has meaning for the resulting *product*, namely the *source code*.

Moreover, the three specified product attributes are quite different and should be stated as separate requirements (preferably in measurable form as discussed below in Section 4).

> *7.4.2.5*: "The design method chosen shall possess features that facilitate software modification. Such features include modularity, information hiding and encapsulation

Although this requirement refers to two *processes* (*design* and *modification*) its primary focus is a *resource*, namely the *design method*. Three very different attributes of the method are specified. The reference to *modification* is out of place here, since the specified properties are only conjectured to be beneficial when subsequent modifications take place.

> *7.4.7.1*: Each module shall be tested against its Test Specification

This is strictly speaking a combination of two separate requirements (and should be treated as such). One is a product requirement: the existence of a document (Software Module Test Specification) to accompany each module. The other is a process requirement that specifies that a certain type of testing activity has to be carried out. The following requirement also says something about the testing process, but is driven by much more specific properties of the product (and hence we would classify it as a product requirement):

> *7.7.2.6b*: "The software shall be exercised by simulation of i) input signals present during normal operation, ii) anticipated occurrences, and iii) undesired conditions requiring system action."

The above classification of standards' requirements represents only the first stage in our proposed means of interpreting standards. It is important because it forces us to identify the specific object of the requirement, and to naturally seek clarification where this is unclear. As a final example, consider the following requirement:

> *7.4.2.8*: "Where the software is to implement both safety and non-safety functions then all of the software shall be treated as safety-related unless adequate independence between the functions can be demonstrated in the design".

By thinking about our classification we can interpret this rather vague and confusing requirement. First of all we tease out the fact that this is a product requirement, but that there are two levels of product being considered: the software as a whole; and the set of individual functions which are being implemented. We need to break up the requirement into the following sub-requirements:

1. The individual functions in the software shall be identified and listed in the Software Architecture Specification; safety-related functions shall be marked as such. (This is a product requirement; the product is the Software Architecture Specification.)

2. An independence check will be performed on each <safety, non-safety> pair of functions identified in (1). (This is a process requirement: how the check is to be performed needs to be further expanded.)

3. A <safety, non-safety> pair of functions are defined to be independent provided that ... (needs to be further expanded) . Two functions that are not independent are defined to be dependent. (Product requirement.)

4. The whole system shall be partitioned into two groups of functions: Group A will contain all safety-related functions together with all non-safety related functions which are dependent on at least one safety related function. Group B will contain all remaining non-safety-related functions. The whole software system shall be classified as safety-related if Group B is empty. (Product requirement.)

## 3.2 Internal and external attributes

For product requirements, we make a distinction between attributes which are internal and those which are external. An *internal attribute* of product X is one that is dependent only on product X itself (and hence not on any other entity, be it another product, process or resource). For example, where X is source code, *size* is an internal attribute. An *external attribute* of a product X is one that is dependent on some entities other than just product X itself. For example, if product X is source code then the *reliability* of X is an external attribute. Reliability of X cannot be determined by looking only at X; it is dependent on the machine and compiler running X, the person using X, and the mode of use. If any of these are changed then the reliability of X can change. We have already seen numerous examples of external attributes in the above requirements (testability, maintainability, readability). Attributes like *modularity* (in 7.4.5.3) can, with specific definitions, be regarded as internal [Fenton and Pfleeger 1996].

The distinction between internal and external attributes is now a widely accepted basis for software evaluation. Clearly, external attributes are the ones of primary concern, especially as our ultimate objective here is to determine acceptance criteria for safety critical systems. This means that we have to determine whether the system's external attributes like safety, reliability, and maintainability are acceptable for the system's purpose. In practice, these attributes cannot be measured directly. We may be forced to make a decision about the acceptability of these attributes before the system is even extensively tested. This means that we are forced to look for evidence in terms of internal product attributes, or process and resource attributes. Requirements in standards which simply state that certain desirable external attributes should be present are invariably vacuous and should be removed (since they are nothing more than *objectives*).

## 3.3. Balance between types of requirements

The Oxford Encyclopaedic English Dictionary defines a standard as "an object or quality of measure derived as a basis or example or principle to which others conform or should conform or by which the accuracy or quality of others is judged"

This definition conforms to the widely held intuitive view that standards should focus on specifying measurable quality requirements of products. Indeed, this is the emphasis in traditional engineering standards. This point was discussed in depth in [Fenton et al 1993] which looked at specific safety standards for products (such as pushchairs). These explicitly specify tests for assessing the safety of the products. That is, they provide requirements for an external attribute of the final product. The measurable criteria for the testing process are also specified. There is therefore a direct link between conformance to the standard and the notions of quality and safety in the final product. Standards such as BS4792 [BSI 1984] also specify a number of requirements for internal attributes of the final product, but only where there is a clearly understood relationship between these and the external attribute of safety.

We contrast this approach with software safety standards. Very few requirements in these standards are well-defined product requirements. For example, [Fenton: et al 1993] provided a detailed comparison of the requirements in BS 4792 with those of DEF-STAN 00-55. The latter consists primarily of process requirements (88 out of a total 115 with 14 internal product and 13 resource requirements. There is not a single external product requirement. In contrast, BS 4792 consists entirely of product requirements (28 in total) of which 11 are external.

The distribution of requirements in 00-55 seems fairly typical of software standards studied in SMARTIE. The standard IEC 1508 is slightly different in that there is a very large number of resource requirements, but again we find far more process than product requirements. The difference between requirements in standards such as 00-55 and IEC 1508 compared with those in BS 4792 is that, generally, there is no conclusive evidence that satisfying them will help achieve the intended aim of safer systems. For example, the following are typical internal product requirements from IEC 1508:

*7.4.4.6*: "The coding standards shall specify good programming practice, proscribe unsafe language features and describe procedures for source code documentation."

*7.4.2.11*: "The software design shall include, commensurate with the required safety integrity level, self-monitoring of control flow and data movements."

*7.4.3.2a*: "The Software Architecture Design Description shall be based on a partitioning into components/subsystems, each having an associated Design Specification and Test Specification."

*7.4.5.3*: "The software should be produced to achieve modularity, testability and maintainability."

Each of these (which would need further clarification to be usable anyway) represent particular viewpoints about internal structural properties that may impact on system safety. Unfortunately, there is no clear evidence that any of them really do [Fenton et al 1994]. The many process and resource requirements in standards such as IEC 1508 have an even more tenuous link with final system safety.

# 4 Classifying standards' requirements by level of objectivity

The above classification of standards' requirements into process, product, or resource represents only the first stage in interpreting standards. The next stage is to further classify the requirements according to the ease with which we can assess conformance. Our objective is to identify the 'rogue' requirements. These are the requirements for which the assessor's obligation (as discussed in Section 2) is unclear; that is, where an assessment of conformance has to be purely subjective. Assuming that a requirement refers to some specific, well-defined process, product or resource, we distinguish four degrees of clarity for each requirement (as shown in Table 1

| Code | Interpretation |
|---|---|
| R | A reference only with no indication of any particular attribute(s) which that entity should possess |
| * | A reference for which only a subjective measure of conformance is possible |
| ** | A reference for which a partially subjective and partially objective measure of conformance is possible |
| *** | A reference for which a totally objective measure of conformance is possible. |

Table 1. Codes for degree of detail given in any requirement

Ideally, the vast majority of requirements should be in categories '**' and '***' (with a small number of necessary 'R's for definition). In the IEE pushchair safety standard BS4792 every one of the 28 requirements is in category '***'. Although IEC 1508 is more objective than the vast majority of software standards reviewed during SMARTIE (and is indeed a significant improvement on its earlier draft IEC SC65A), many requirements (including most of the examples presented so far) still fall into the 'R' and '*' category. This means that conformance to such requirements can only be assessed purely subjectively. It is difficult to justify their inclusion in a safety critical standard. How are we to assess, for example, requirements such as:

*7.4.6.1a*: "The source code shall be readable, understandable, and testable"

It would be near impossible to convincingly whether it is satisfied or not, so it is effectively redundant. Alternatively, we could attempt to re-write it in a form which enables us to check conformance objectively. As long as there is mutual agreement (between developer and assessor) in the overall value of a requirement (however vague) then this is the option we propose. First of all, we stress that there is a considerable difference between

a) making a requirement objective, and

b) being able to assess conformance to a requirement objectively.

Option (a) is generally very difficult and often impossible; in an immature discipline there is even some justification for allowing a level of subjectivity in the requirements. It is only option (b) that is being specifically recommended. The following example explains the key difference between (a) and (b) and shows the different ways we might interpret requirements to achieve (b). There are generally many ways in which this can be done:

*Example 1*: We consider how we might interpret requirement 7.4.6.1a above in order that we can assess conformance objectively. First of all we note that there are actually three separate product requirements, namely:

i) Each software module shall be readable;
ii) Each software module shall be understandable;
iii) Each software module shall be testable.

We concentrate on just (i) here. Consider the following alternative versions:

A. To accompany each software module a report justifying readability will be produced.
B. To accompany each software module a report justifying readability will be produced. This report will include a section that explains how each variable in the module is named after the real-world entity that it represents.
C. The ratio of commented to non-commented code in each software module shall be at least 1 to 4, and the total size shall not exceed 1000 LOC.
D. An independent reviewer, with a degree in Computer Science and 5 years experience of technical editing, shall devote a minimum of 3 hours to reviewing the code in each software module. The reviewer shall then rate the module for readability on the following 5-point ordinal scale: 0 (totally unreadable) 1 (some serious readability problems detected, requiring significant re-write); 2 (only minor readability problems detected, requiring re-write); 3 (only trivial readability problems detected); 4 (acceptable). The module must achieve a rating of 3 or higher.

Each of the above versions can be checked for conformance in a purely objective manner even though a large amount of subjectivity is still implicit in each of the requirements. In the case of A we have only to check the existence of a specific document. This is a trivial change to the original requirement since we have still said nothing about how to assess whether the document adequately justifies whether the module is readable. Nevertheless we have pushed this responsibility firmly onto the developers and not the assessors. Alternative B is a refinement of A in which we identify some specific criteria that must be present in the document (and which might increase our confidence in the readability argument). For alternative C we have only to check that the module has the right 'measures'. A simple static analysis tool can do this. In alternative D we have only to check that the rating given by the independent reviewer is a 3 or 4 and check that this person does indeed have the specified qualifications and experience.

In each of the alternative versions measurement plays a key, but very simple role. In the case of version D the requirement is based on a very subjective rating measure. Nevertheless we can determine conformance to this requirement purely objectively.

None of the alternative requirements except C is a requirement for which the module itself (a product) is the focus. Alternatives A and B are both requirements of a different product, while Alternative D concentrates on the results of a reviewing process .

Example 1 confirms that being able to assess conformance to a requirement objectively does not mean that the requirement itself is objective. Nor, unfortunately, does it always mean that assessment will be easy. The approach that we are proposing is to move toward identifying measurable criteria to replace ill-defined or subjective criteria. This is consistent with the traditional measurement-based approach of classical engineering disciplines. Texts such as [Fenton and Pfleeger 1996] explain how to move toward quantification of many of the subjective criteria appearing in a standard such as IEC 1508. The following example further illustrates the method:

*Example 2*: Requirement 7.2.2.5a asserts "To the extent required by the integrity level the Software Safety Requirements Specification shall be expressed and structured in such a way that it is as clear, precise, unequivocal, verifiable, testable, maintainable and feasible as possible commensurate with the safety integrity level". Each of the required attributes here (which need to be treated as separate requirements) are ill-defined or subjective. In the case of 'maintainable' there are a number of ways we could interpret this so that we could assess conformance objectively. The most direct way is to specify a mean or maximum time in which a change to the SSRS can be made. Since such measures are hard to obtain it may be preferable to specify certain internal attributes of the SSRS, such as: the electronic medium in which it must be represented; the language in which it has to be written; that it has to be broken up into separately identifiable functions specified using less than 1000 words each; etc. Specification measures such as Albrecht's Function Points [Albrecht 1979] might even be used. A radically different approach is that of alternative D in Example 1 where we simply specify what expert's rating of maintainability has to be achieved.

## 5 Measurement Based Standards Evaluation

So far we have concentrated on how we can interpret and use standards despite their many weaknesses. We do not question the general importance and value of standards to safety critical systems development. Nevertheless, there are very wide differences of emphasis in specific safety-critical standards. For example, 00-55 and IEC1508 are totally different in their underlying assumptions about what constitutes a good software process; 00-55 mandates the use of formal specification (and is structured around the assumption that formal methods are used), while 1508 mentions it only as a technique which is 'highly recommended' at the highest safety integrity level (level 4). Clearly the standards cannot all be equally effective. They are certainly not equally easy to apply or assess. Therefore we have to assume that some standards are better than others and some requirements in standards are more important than others. Unfortunately, *a priori* we do not know which.

It follows that there is a need for assessing the effectiveness of standards, especially when we consider the massive technological investments which may be necessary to implement them. What we have described so far may be viewed as a 'front-end' procedure for standards evaluation. This is like an intuitive quality audit, necessary to establish whether a given standard satisfies some basic criteria. It also enables us to interpret the standard, identify its scope, and check the ease with which it can really be applied and checked. However, for proper evaluation we need to demonstrate that, when strictly adhered to, the use of a standard is likely to deliver reliable and safe systems at an acceptable cost.

The SMARTIE project looked at how to assess standards in this respect [Pfleeger et al 1994]. The basic impediment to proper evaluation is the sheer flabbiness of the relevant standards. Many of the standards address the entire development and testing life-cycle, containing massive (and extremely diverse) sets of requirements. It makes no scientific sense, and is in any case impractical, to assess the effectiveness of such large objects. Thus we use the notion of a *mini-standard*. Any set of requirements, all of which relate to the same specific entity or have the same specific objective, can be thought of as a standard in its own right, or a mini-standard. Rather than assess an entire set of possibly disparate requirements, we instead concentrate on mini-standards.

The need to decompose standards into manageable mini-standards is a key stage in the evaluation procedure described in [Pfleeger et al 1994]. Many software-related standards are written in a way which makes this decomposition extremely difficult. However, the software part of IEC 1508 is structured in a naturally decomposable way. We can identify seven key mini-standards in the relevant parts of IEC 1508:

1. Process of Specifying Safety Integrity Levels (Part 1, Section 8)
2. The Safety Plan (Part 1, Sections 6 and 7, and Part 3, Section 7.1 and 7.2)
3. Resources (Part 1, Section 5 and Part 3 Section 5 which concentrate on people; and those requirements in Part 3 which describe the requirements of the design method, programming language and other tools)
4. The Software Requirements Specification (Part 3, Section 7.2)
5. The Design Process (Part 3, Section 7.4)
6. The validation and verification and testing process (Part 3, Sections 7.3, 7.7 and 7.9)
7. The maintenance process (Part 3, Sections 7.6 and 7.8)

The formal obligations for evaluating the efficacy of a mini-standard reduces to measuring the following criteria in a given application of the standard:-

- *Benefits*: What observable benefits are supposed to result from the application of the mini-standard? Specifically, which external attributes of which products, processes or resources are to be improved? For example, is it reliability of code, maintainability of code or designs, productivity of personnel?

- *Degree of conformance:* To what extent have the requirements been conformed to (note that to measure this properly we need to be able to assess conformance objectively)
- *Cost:* What is the cost of applying the mini-standard (over and above normal development costs).

Essentially, a mini-standard successfully passes an evaluation for a specific environment if, in such an environment, it can be shown that the greater the degree of conformance to the standard, the greater are the benefits, providing that such improvements merit the costs of applying the standard.

The problem of over-emphasis on process requirements in safety-critical standards has an important ramification when it comes to the evaluation procedure. Specifically, we found that, for many process requirements, the intended link to a specific benefit is unclear. For example, 00-55 contains the requirement:

> *30.1.2:* "The Design Authority shall use a suitable established and standardised Formal Method or Methods for the Formal Design. Properties that cannot be expressed using the Formal Method or Methods shall be notified to the MOD(PE) PM and a suitable, established design method agreed."

Even if we could determine objectively conformance to such a requirement—the appendix of the standard provides some crude guidelines for this—it is unclear what the specific intended benefit is. Only from reading the rest of the standard do we discover that an intended major benefit is that it helps to make implementations 'provable' (that is it makes possible a mathematical proof of correctness). However, this in itself would be of little interest as a benefit to users. Rather, we have to assume the implicit benefit to be *implemented code which is more reliable.*

# 6 Summary and Conclusions

For safety-critical standards to be usable we expect the individual requirements to be clear to:

> *Developers* so that they know what they are required to do; and

> *Assessors* so that they know how to determine conformance.

Unfortunately, many requirements in the relevant standards are not clear in either of these respects (although IEC 1508 shows a significant improvement on its previous incarnation IEC SC65A in many of the specific respects identified here). We have shown how to interpret unclear requirements in both respects, but with special emphasis on the assessors' needs. There is a significant different between:

a) making a requirement objective, and
b) being able to assess conformance to a requirement objectively.

While (a) is generally very difficult, we have shown how to achieve (b) in a rigorous manner.

The vast majority of all requirements in existing safety-critical systems standards are unnecessarily unclear. Our approach to interpreting such requirements begins by teasing out the relevant process, product or resource that is the primary focus. In many cases this means breaking down the requirement into a number of parts. This technique alone can often achieve the required level of clarity. We provided numerous examples drawn from IEC 1508 on how to do this.

When the requirements in safety critical standards are classified according to products, processes and resources, we found a dearth of external product requirements (in stark comparison with safety-related standards in traditional engineering disciplines). The emphasis was on process and resource requirements with a smaller number of internal product requirements. This balance seems inappropriate for standards whose primary objectives are to deliver products with specific external attributes, namely safety and reliability.

Finally, we discussed the need for assessing the effectiveness of standards. The sheer size of existing standards makes them too large to assess as coherent objects. Thus we used the notion of mini-standards, whereby we identify coherent subsets of requirements all relating to the same specific process, product, or resource. The identification of mini-standards helps us not only in assessment but also in rationalising and interpreting standards. We proposed a decomposition of IEC 1508 into mini-standards.

We have presented some simple practical advice on how to improve safety-related standards. Unfortunately, the standards-making process is long and tortuous. In many cases this process itself contributes to some of the problems highlighted earlier. Perhaps it is time that the software industry paid for the development of good, timely standards rather than continued to rely on the contributions of individuals who volunteer their effort to standards' making bodies. While such contributions are, more often than not, heroic and unsung, they are nevertheless entirely ad-hoc. A such we deserve nothing better than the ad-hoc standards we have at present.

# 7 Acknowledgements

The contents of this report have been influenced by material from the SMARTIE project (funded by EPSRC and DTI) in which the author was involved, and also by an earlier assessment of IEC SC65A that the author performed as part of the ESPRIT project CASCADE project (funded by the CEC). The new work carried out here was partly funded by the ESPRIT projects SERENE and DEVA. The author is indebted to Colum Devine, Miloudi El Koursi, Simon Hughes, Heinrich Krebs, Bev Littlewood, Martin Neil, Swapan Mitra, Stella Page, Shari Lawrence Pfleeger, Linda Shackleton, Roger Shaw and Jenny Thornton for comments that have influenced this work.

# 8 References


Albrecht A.J, Measuring Application Development, Proceedings of IBM Applications Development joint SHARE/GUIDE symposium. Monterey CA, pp 83-92, 1979.

British Standards Institute, Specification for Safety Requirements for Pushchairs, British Standards Institute BS 4792, 1984.

Fenton NE and Pfleeger SL, Software Metrics: A Rigorous and Practical Approach (2nd Edition), International Thomson Computer Press, 1996.

Fenton NE, Littlewood B, and Page S, Evaluating software engineering standards and methods, in Software Engineering: A European Perspective (Ed: Thayer R, McGettrick AD), IEEE Computer Society Press, pp 463--470, 1993.

Fenton NE, Pfleeger SL, Glass R, Science and Substance: A Challenge to Software Engineers, IEEE Software, 11(4), 86-95, July, 1994.

IEC (International Electrotechnical Commission), Software for computers in the application of industrial safety related systems, IEC 65A, 1992.

IEC (International Electrotechnical Commission), Functional safety of electrical/electronic/programmable systems: generic aspects, IEC 1508, 1995.

Ministry of Defence Directorate of Standardization, Interim Defence Standard 00-55: The procurement of safety critical software in defence equipment; Parts 1-2, Kentigern House 65 Brown Street Glasgow, G2 8EX, UK, 1991.

Pfleeger SL, Fenton NE, Page P, Evaluating software engineering standards, IEEE Computer, 27(9), 71-79, Sept, 1994.

# Engineering Cognitive Diversity

S.J. Westerman, N.M. Shryane, C.M. Crawshaw, & G.R.J. Hockey
Department of Psychology, University of Hull, Hull, HU6 7RX, England.

## Abstract

This paper discusses the potential advantages of developing cognitive diversity within the process of safety-critical systems design. Two broad approaches to achieving diversity are identified. The first requires diversity to be created within the task environment. The second relies on individual differences in task performance that can be used to engineer diversity. Empirical evidence is presented to support the potential of both methods, and implications for future progress in this field are discussed.

# 1 The need for cognitive diversity

A widely recognised method of improving system reliability is to use redundancy of components. In the event of component failure, system failure is averted because another component is available to continue functioning. Although this approach can be very effective for system hardware, redundancy within the context of human system components (e.g. n-version software, independent testing) is not sufficient to ensure reliability [Senders 91]. System designers are prone to 'common mode' errors, in which two or more individuals make exactly the same error of cognition. For example, it has been experimentally demonstrated that independently coded n-version software does not fail independently [Leveson 95]. It has been argued that cognitive diversity presents a potential means of improving this situation by reducing the probability of common mode human error [Westerman 95].

The term 'cognitive diversity' is preferred, in this context, to the term 'design diversity', that has been used in some previous research [Avizienis 84] [Lyu 92] for two reasons. First, the term explicitly recognises the pivotal importance of human cognition within the process of achieving diversity. Even when the benefits of diversity have been established in previous studies, for example using multiple languages for n-version software development [Lyu 92], the cognitive mechanisms responsible for diverse behaviour remain unchallenged. An understanding of these mechanisms is required if diversity is to be effectively harnessed as a method of improving system reliability. Second, the principles involved extend beyond the design process and may be applicable to system operation and maintenance.

A number of assumptions must be satisfied before cognitive diversity can make a contribution to fault tolerance [Westerman 95].

*Assumption 1 : Individuals are not capable of consistently achieving a perfect solution.* If the difficulty of a task is such that individuals are consistently capable of error free performance then there would be no need to devise a diverse system.

*Assumption 2 : There is some 'constancy' in error.* There must be an association between the circumstances that will promote error and the nature of that error (see [Reason 90]).

*Assumption 3 : The errors made by one individual or strategy are not a superset of the errors made by another individual or strategy.* If this were the case, the performance of one individual/strategy could be said to be better, but there would be no diversity.

*Assumption 4 : Diversity can be operationalised in a psychologically meaningful way.* Although it may be possible to determine a number of diverse solutions to a problem, and to implement diverse solutions within computer-based expert systems, e.g. [Lee 93], unless each component represents a psychologically meaningful entity cognitive diversity will not be possible.

It follows that if these assumptions are satisfied, and cognitive diversity is achieved, this will render task performance less susceptible to common mode error. It should be noted, however, that there are a number of organisational and psychological factors that militate against cognitive diversity. For example, organisational staff selection and training programs tend to produce conformity of cognition between individuals. Cognitive diversity must be actively engineered.

It is possible to identify two broad approaches to the development of cognitively diverse human system components. First, it may be possible to engineer diversity within the task environment in order to encourage cognitive diversity. Within this context the task environment can be defined as the specified methods, equipment (hardware and software), and personnel with which the individual must interact in order to perform a given task. Second, it is possible that diversity can be more directly engineered on the basis of psychometric assessments of human system components. Individuals will vary with respect to their mental model of the task, their abilities, and their preferred task performance strategies. If these characteristics are associated with the detection of specific error types (Assumption 2) then they can be used as a basis for cognitively diverse systems. This paper considers both of these approaches within the context of safety-critical system verification. In the following sections some literature bearing on these issues is reviewed, and some empirical evidence from our laboratory is presented.

## 2 Manipulating the task environment

It can be hypothesised that task environments vary in the cognitive demands they place on engineers, such that, for any given task environment, some specific types of error are detected more readily than others  (Assumption 2). Upon this basis, a number of different approaches to the creation of diverse task environments are possible [Avizienis 95], e.g. using different programming languages; software development methods; programming tools and environments. However, there is little empirical evidence that bears directly upon this issue.

Our interest in task environment diversity stems from a field study examining the safety critical design task of verifying the geographic data that controls the safe movement of trains within sections of railway. The design process is such that, once geographical data has been written, engineers perform an independent code inspection, and this is followed by an independent functional test (see [Westerman 94] for more details). Our field study indicated that, when errors were classified according to the task related principles violated (e.g. allowing opposing routes to be set and thereby creating the conditions for a collision between trains), code inspection and functional testing design processes were detecting different types of error [Westerman 95]. However, on the basis of this data it was not possible to determine whether diversity was truly present because: i) testing was always preceded by code inspection and consequently there was limited opportunity for this process to detect those types of errors that code inspection was particularly efficient at detecting; and ii) not all errors were equally visible to both methods. For example, errors in the layout of data were not visible to an engineer performing the functional test. On the basis of this evidence, it is possible that testing locates a superset of those errors that are equally visible to both methods (see Assumption 3, above). Many previous research efforts suffer from similar shortcomings, and address the overall efficiency of different methods but not the degree of diversity (although see [Basili 87] [Shimeall 91])

In order to pursue this issue two laboratory experiments were conducted, using a simulation of the geographic data verification task (the primary differences between these experiments are described in Section 3, below). Participants were recruited from the student populations of the Universities of Hull or Bradford. Participants for Experiment I had a computer science or engineering background, participants for Experiment II had a science background. In each experiment participants were required to verify geographic data on the basis of either a code inspection or a functional test. A number of faults were introduced into the data that were designed to contravene one of four signalling principles (the rules governing system behaviour). Faults were further classified according to whether they resulted from errors of omission or commission in the data. Both experimental conditions (code inspection and functional testing) used the same geographic data (including the same faults). All faults were selected so that they would be 'visible' to either method of detection (code inspection or functional test).

The main effect of task environment (code inspection vs functional test) on error detection performance was not significant for either experiment. However, in both cases there was a significant interaction between task environment, and error type such that task performance methods were differentially sensitive to different error types. As can be seen from Table 1, in Experiment I, testing appeared to be better than code inspection at detecting errors relating to setting an opposing route over a different point lie, but was comparatively poor at detecting OPT errors (ensuring that track circuits were clear before showing a proceed aspect at a signal).

Table 1. Experiment I: Proportion of faults detected by task environment, signalling principle, and omission vs commission.

| | Code inspection | | Functional test | |
|---|---|---|---|---|
| | Omission | Commission | Omission | Commission |
| ORDL | .667 | .708 | 1.000 | 1.000 |
| ORSL | .917 | .750 | .929 | .857 |
| OPT | .958 | .917 | .857 | .405 |
| OTHER | .958 | .833 | .893 | 1.000 |

*Note: ORDL = opposing route with different points lie; ORSL = opposing route with same points lie; OPT = output file failing to make sure that all sections of track are clear before the signal shows a proceed aspect; and OTHER = miscellaneous route setting errors not included in the above.*

Overall error detection was poorer in Experiment II (see Table 2). This may be attributable to less stringent sampling (see above). However, consistent with Experiment I, code inspection detected fewer faults in setting an opposing route over a different lie of points that resulted from an omission in the data; and testing was comparatively poor at detecting OPT errors. Testing was also better at detecting faults in setting and opposing route over the same lie of points. This latter finding is consistent with the results of the previously mentioned field study.

Table 2. Experiment II: Proportion of faults detected by task environment, signalling principle, and omission vs commission.

| | Code inspection | | Functional test | |
|---|---|---|---|---|
| | Omission | Commission | Omission | Commission |
| ORDL | .396 | .583 | .709 | .250 |
| ORSL | .313 | .229 | .459 | .584 |
| OPT | .750 | .750 | .584 | .375 |
| OTHER | .875 | .729 | .667 | .646 |

*Note: ORDL = opposing route with different points lie; ORSL = opposing route with same points lie; OPT = output file failing to make sure that all sections of track are clear before the signal shows a proceed aspect; and OTHER = miscellaneous route setting errors not included in the above.*

These results provide evidence of cognitive diversity between code inspection and functional testing. This may be attributable to the types of cognitive processing that is required to verify correctness for each of the error types. It can be hypothesised that, in the case of opposing routes, the functional representation of the railway used in the testing task environment assists visualisation of test cases and thereby reduces processing demands. In contrast, attempting to detect OPT errors requires that the state of each track circuit is examined, and this appears more laborious in the testing condition than in the code inspection condition. This hypothesis might be tested in a further experiment by isolating each of these fault types and examining performance in each task environment in relation to the mental workload experienced by participants.

However, two broader models of cognition may be applied to these results. First, it would seem that these task environments differ with respect to the level of abstraction with which information is presented to participants. Rasmussen [Rasmussen 86] proposed that systems can be represented at various levels of abstraction, in which high levels are concerned with broad system function and lower levels are concerned with the 'nuts and bolts' of the system. In this respect, functional testing can be seen to represent a higher level of system abstraction code inspection. An alternative model for diversity in this instance concerns the relative spatial vs verbal processing demands of the task environments. It can be hypothesised that functional testing provides greater support for spatial processing of the task, whereas verbal processing is emphasised in the case of code inspection. Further experimentation is required to determine whether either, or both, of these models are appropriate.

# 3 Manipulating individual differences

A second approach to achieving cognitive diversity in human system components is to utilise differences that are specific to the individual rather than the task environment. Independence between stages/personnel is commonly used in the hope that the influence of randomly occurring individual differences will be encouraged and that diversity will result. However, there are also examples of design processes where more formal recognition is given to the role of individual differences, e.g. Fagan's inspection teams that comprise individuals with different task roles [Fagan 76] [Fagan 86].

Although it is possible that, if particular task performance strategies could be identified as achieving diverse solutions, individuals could be trained to perform a task in different ways, the evidence regarding the application of strategy differences in isolation from task environment differences is not overly promising. This paper will focus on existing individual differences and their influence on error detection performance.

## 3.1 Cognitive ability

There are a number of dimensions of individual difference that might predict diverse error detection performance, e.g. personality, cognitive ability, domain knowledge, expertise. As an initial foray in this area we examined the influence of spatial and verbal ability. Although there is a substantial body of evidence supporting an association between cognitive ability and programming skills (e.g. see [Egan 88]), studies addressing this issue have been concerned with predictors of absolute levels of performance. When considering cognitively diverse systems of program verification, what is of interest is the association between diversity of individual characteristics, on the one hand, and diversity in error detection, on the other.

A fundamental distinction has been proposed in the nature of mental representation used by individuals during problem solving. It has been suggested that some individuals, with high spatial ability, will tend to use a mental representation that is essential spatial, whilst other individuals, with high verbal ability, will tend to use a mental representation of the task that is essentially verbal (e.g. see [Hunt 78]). It can be hypothesised that such individual differences will form the basis of cognitive diversity, in that some types of error will be more readily detected using a spatial representation of the problem domain, while other error types will be more readily detected using a verbal representation of the problem domain.

In order to test this hypothesis, as part of Experiment I described above, participants also completed psychometric tests of spatial and verbal ability. A measure of relative spatial/verbal ability was derived for each individual by subtracting standardised test scores for verbal ability from standardised scores for spatial ability (see [Cronbach 81]). Performance of each possible code inspection/functional testing pairing of participants was examined. For each pair the difference in relative ability scores was calculated, along with a measure of error detection performance. For each specific error, if one or both members of the pair detected that specific error the error detection score for a pair was incremented by one. The difference in ability scores for the pair was then regressed on this error detection measure. No significant association was found. It would appear that differences in spatial/verbal ability are not associated with diverse error detection performance.

It is possible that this non-significant result was due to a biased sample. Recruiting participants with an engineering background will tend to result in sample spatial ability scores that are higher than those found in the general population. It may be, therefore, that all participants tended to use a spatial, as opposed to verbal, mental representation of the task. This would be difficult to test. However, it can be argued that similar constraints would apply in 'real world' settings, and that for a demonstration of differences to be meaningful, this must be achieved with a representative sample.

## 3.2 Mental models

A more direct approach to predicting diversity within individuals is to examine the mental model that individuals apply to task performance. A mental model is "...a rich and elaborate structure, reflecting the user's understanding of what the system contains, how it works, and why it works that way" [Carroll 88]. It should not necessarily be regarded as an accurate view of the world [Norman 86], and for this reason knowledge of the mental model that an engineer applies to a task may enable prediction of patterns of error. However, the difficulty with this approach lies in achieving a quantitative assessment of a mental model. One possible method is to use psychometric measures of conceptual distance with respect to concepts within the task environment (see [Cooke 94] for review). Although little work has been done in this area, there are some encouraging indications of the validity of this method [Coury 92] [Federico 95] [Pallant 96].

In order to test the hypothesis that cognitive diversity could be predicted in this way, participants in Experiment II, described above, also completed a number of psychometric tests relating to their mental model of the task environment. This followed an initial training period that was common to both code inspection and functional testing conditions. Concepts used in this psychometric assessment were combinations of the physical components of the railway (points, signals, routes, and track circuits) and their possible binary states (e.g. free vs locked). Three questionnaires were administered, as follows:

1. An assessment of conceptual similarity (using a 0 to 100 rating scale) for each possible pairing of concepts.
2. An assessment for each possible pairing of concepts (using a -100 to +100 rating scale) which was the most strongly related to the further concept of 'system safety'.
3. As for item 2 above but with respect to the further concept of 'system functionality'.

A two-dimensional multidimensional scaling solution was derived from each of these questionnaires. INDSCAL [Norusis 92] was then used to examine individual differences, with flattened weights providing an index of individual deviations from the mean solution. As with the analysis of cognitive ability data, performance was examined for all possible pairs of code inspection / functional testing participants. Difference scores for the pair (taken to be an index of mental model dissimilarity) for each psychometric measure were regressed onto error detection performance for the pair (calculated as described above).

The regression equation accounted for significant variance, $F(1,142) = 8.86$, $p<.005$. Mult. $R = .24$, R Sq. $= .05$, with functionality questionnaire scores (questionnaire 3 above) being significantly positively associated with performance, Beta $= .24$, $t = 2.98$, $p<.005$. When difficult errors (the 50% of errors that were least frequently detected) were considered in isolation slightly more variance could be accounted for, Mult $R = .29$, R Sq. $= .08$, $F(1,142) = 12.80$, $p<.001$. The functionality questionnaire was, again, the only significant predictor, Beta $= .29$, $t = 3.58$, $p<.001$.

Although the amount of variance in error detection performance accounted for by the psychometric measure used in this study was comparatively small, the results indicate a significant association between mental model diversity and error detection performance. Within the context of a safety-critical design task even small improvements in reliability may be important. Further, it should be noted that the error detection variance accounted for is in addition to the benefits, described earlier, associated with combining checking and testing task environments. Although these results must be considered preliminary they are encouraging. A technique such as this would be well suited to applied contexts where, for a small outlay in terms of time and materials, it would appear that improvements in reliability can be made.

# 4 Discussion

The empirical studies of a system verification task, presented above, indicate that cognitive diversity can be achieved by both of the approaches described in Section 1 above. It was apparent that the use of diverse task environments, in the form of code inspection and functional testing, resulted in greater diversity in error detection performance, with each of the four assumptions, detailed in the introduction, being satisfied. However, it is not clear what precise cognitive principles were involved in achieving diversity within this setting and further research is required.

The use of psychometric measures of mental model diversity, in Experiment II, enabled additional variance in error detection performance to be predicted. However, these results require replication and extension. There are a number of important questions to be addressed. For example: are there better ways of probing mental models? will the same method be successful with task experts? are these results replicable in different task situations? how stable are these psychometric assessments over time? In contrast, individual differences in spatial and verbal ability did not appear to predict diverse error detection. It would seem that, when performing a system verification task, individuals who can be characterised as high spatial do not differ in the type of errors they detect from those who could be characterised as high verbal.

It is important to recognise that, although this paper has considered two approaches to cognitively diverse task performance in isolation, the nature of the task environment may interact with characteristics of the individual. For example, it has been demonstrated [Adelson 84] that the performance of experts was better than that of novices when presented with materials encouraging an abstract representation of a programming task. However, when materials encouraging a concrete representation were presented the performance of novices surpassed that of experts. Further, on the basis of a study of program debugging, Vessey concluded that "...experts are situation dependent problem solvers, while novices are situation independent problem solvers" [Vessey 1986].

It is early days for the investigation of cognitive diversity, but the evidence presented above suggests that it may be a fruitful area for psychological inquiry. A taxonomy of diversity is required that addresses cognitive underpinnings of task performance, that includes estimates of potential reliability gains, and that provides an assessment of the difficulty of instituting diverse systems. If diverse methods (such as the quantitative appraisal of mental models) are relatively easy to achieve, even modest improvements in reliability make these methods worthy of application. If methods account for more substantial variance in performance, then even if they require substantial task restructuring then they will be worthwhile. It seems that reliability improvements can be made by engineering cognitive diversity.

# References

[Adelson 84]     Adelson, B. "When novices surpass experts: The difficulty of a task may increase with expertise". Journal of Experimental Psychology: Learning, Memory, and Cognition, 1984;10: 483-495.

[Avizienis 84]     Avizienis, A.A. & Kelly, J.P.J. "Fault tolerance by design diversity: Concepts and experiments". IEEE Computer, 1984; 17: 67-80.

[Avizienis 95]	Avizienis, A.A.. "The methodology of N-version programming". In M.R. Lyu (Ed.), Software Fault Tolerance. Wiley, 1995. pp. 23-46.

[Basili 87]	Basili, V.R. & Selby, R.W. "Comparing the effectiveness of software testing strategies". IEEE Transactions on Software Engineering, 1987; 13: 1278-1296.

[Carroll 88]	Carroll, J.M. & Olson, J.R. "Mental models and human-computer interaction". In M. Hellander (Ed.) Handbook of Human-Computer Interaction. North Holland, 1988. pp. 45-65.

[Cooke 94]	Cooke, N.J. "Varieties of knowledge elicitation techniques". International Journal of Human-Computer Studies, 1994; 41: 801-849.

[Coury 92]	Coury, B.G., Weiland, M.Z., & Cuqlock-Knopp, V.G. "Probing the mental models of system state categories with multidimensional scaling". International Journal of Man-Machine Studies, 1992; 36: 673-696.

[Cronbach 81]	Cronbach, L.J. & Snow, R.E. "Aptitudes and Instructional Methods". Irvington, 1981.

[Egan 88]	Egan, D.E. "Individual differences in human-computer interaction". In M. Hellander (Ed.) Handbook of Human-Computer Interaction. North Holland, 1988. pp. 543-569.

[Fagan 76]	Fagan, M.E. "Design and code inspections to reduce errors in program development". IBM Systems Journal, 1976; 3: 182-211.

[Fagan 86]	Fagan, M.E. "Advances in software inspections". IEEE Transactions on   Software Engineering, 1986; 12: 744-751.

[Federico 95]	Federico, P-A. "Individual differences in metacognitive decision making and situation assessment". In Proceedings of the Human Factors and Ergonomics Society 39th Annual Meeting. HF&ES, 1995. pp. 878-881.

[Hunt 78]	Hunt, E. "Mechanics of verbal ability". Psychological Review, 1978; 85: 109-130.

[Katz 88]	Katz, I.R. & Anderson, J.R.. "Debugging: An analysis of bug location strategies". Human-Computer Interaction, 1988; 3: 351-399.

[Lee 93]	Lee, J.M. & Kim, J.H. "An integration of heuristic and model-based reasoning in fault diagnosis". Engineering Applications of Artificial Intelligence, 1993; 6: 345-356.

[Leveson 95]	Leveson, N.G. "Safeware: System Safety and Computers". Addison-Wesley, 1995.

[Lyu 92]        Lyu, M.R. & Avizienis, A. "Assuring design diversity in N-version software: A design paradigm for N-version programming". In J.F. Mayer & R.D. Schlichting (Eds.) Dependable Computing for Critical Applications 2. Springer-Verlag: New York, 1992. pp. 192-218.

[Norman 86]     Norman, D.A. & Draper, S.W. "Cognitive engineering". In D.A. Norman & S.W. Draper (Eds.) User Centered System Design - New Perspectives on Human-Computer Interaction. Erlbaum, 1986. pp. 31-61.

[Norusis 92]    Norusis, M.J. "SPSS for Windows: Professional Statistics. Release 5". SPSS: Chicago, IL., 1992.

[Pallant 96]    Pallant, A., Timmer, P. & McRae, S. "Cognitive mapping as a tool for requirements capure". In S.A. Robertson (Ed.), Contemporary Ergonomics. Taylor & Francis, 1996. pp. 495-500.

[Rasmussen 86] Rasmussen, J. "Information Processing and Human-Machine Interaction". North Holland. 1986.

[Reason 90]     Reason, J. "Human Error". Cambridge: Cambridge University Press.    1990.

[Senders 91]    Senders, J.W. & Moray, N.P. "Human Error: Cause, Prediction, and Reduction". Hillsdale, NJ: Lawrence Erlbaum. 1991.

[Shimeall 91]    Shimeall, T.J. & Leveson, N.G. "An empirical comparison of software fault tolerance and fault elimination". IEEE Transactions on Software Engineering, 1991; 17: 173-182, 1991.

[Vessey 86]     Vessey, I. "Expertise in debuggin computer programs: An analysis of the content of verbal protocols". IEEE Transactions on Systems, Man, and Cybernetics, 1986; 16: 621-637.

[Westerman 94] Westerman, S.J., Shryane, N.M., Crawshaw, C.M., Hockey, G.R.J., & Wyatt-Millington, C.W. "Task analysis of the solid state interlocking design process". Report No. SCS-01. Department of Psychology, University of Hull, 1994.

[Westerman 95] Westerman, S.J., Shryane, N.M., Crawshaw, C.M., Hockey, G.R.J., & Wyatt-Millington, C.W. "Cognitive Diversity: A structured approach to trapping human error". In G. Rabe (Ed.), SafeComp '95: Proceedings of the 14th International Conference on Computer Safety, Reliability and Security, Belgirate, Italy, 11-13 October, 1995. pp. 142-155.

# The PRICES Approach to Human Error

Christine Tomlinson[1]
Lloyd's Register
London (U.K.)

## Abstract

Lloyd's Register is managing the PRICES[2] project which aims to
provide guidance for software developers on how to develop
systems which are safe, taking into account practical issues such as
productivity of the development team. To ensure the practicality of
the guidance, PRICES undertook a survey of current working
practices for developing safety-critical software; an in-depth field
study; and several case studies. The guidance is being packaged as a
Code of Practice, which has a companion volume providing the
rationale for the advice given. The companion volume divides into
two parts. One part provides supporting information in the form of
more detailed guidance, references to relevant work etc. The second
part provides an overall theoretical justification for the coverage of
the guidance - inclusion of topic was mainly determined by
likelihood of human error affecting productivity and integrity. It will
be apparent to readers of the Code of Practice that PRICES has taken
a broad view of human error, believing that human error can
emanate from diverse points in what is an intellectual, social and
organisational process. PRICES looked to traditional models of
human error to provide a theoretical underpinning for this view but
found them to be too individualistic. Instead PRICES drew from
work done in the fields of knowledge based systems development,
social psychology and education. This paper begins with a review of
the traditional approaches and discusses why they were considered
to be not appropriate for analysing human error in safety-critical
software development. The paper then discusses alternative views of
human error which, it is claimed, are more appropriate for analysing
human error in safety-critical software development.

---

[1] The author of this paper retains the right of subsequent publication of this paper. Any
opinions expressed in statements made in this paper are those of the individual and not
those of Lloyd's Register.

[2] PRICES is an acronym standing for Productivity, Integrity and Capability Enhancement for
Software. The project consortium is: Lloyds Register, Rolls Royce, BAeSEMA, GP-Elliott
Electronic Systems, Analysis International, Open University and City University. Nuclear
Electric and the University of Hertfordshire are also contributing to the Code of Practice.

# 1    Introduction

Models of human error have been utilised by several of the projects in the Safety-Critical Systems Research Programme.   All of these, e.g. the DATUM, FASGEP and HUFIDESAC projects, have based their approaches on Reason's (1990) or Rasmussen's (1983) models.  These are not surprising choices.  Reason's model is a refinement of Rasmussen's, and together the two models underpin virtually all work on human error in working environments.  However, the relevance of these models to the software development process is questionable.  The models were developed to help minimise the occurrence of operators' mistakes in safety-critical operations in potentially hazardous sectors such as Nuclear, Chemical Processing, and Offshore.   They were developed specifically for scenarios of isolated operators manning machines, who were performing vigilance or control tasks which required neither any degree of intellectual activity nor any social interaction - at least in theory.   The limitations of these individualistic, cognitive models are accepted by their proponents:

"The final note is a rather pessimistic one... While cognitive psychology can tell us something about an individual's potential for error, it has very little to say about how these individual tendencies interact within complex groupings of people working in high-risk systems.  And it is these collective failures that represent the major residual hazard." (Reason, 1990 p.xii)

Clearly, the models were never intended for explaining the occurrence of error in intellectual teamwork such as software development, but the appropriateness of the models is critical:

"The design and analysis methods used for safety-critical systems are based on particular underlying models of the accident process and of human errors.  How effective our procedures are depends, to a large extent, on how accurate our models are - that is, how well they reflect the features of the environment to which they are applied." (Leveson 1995, p.185)

The main feature of the software development process which the models should reflect, is that it is a highly creative process requiring the collective effort of specialists with different kinds of expertise.  The project team face the intellectual challenge of developing a shared understanding of what is required and of working out how to accomplish it.  The project has to be managed so that the product is completed on time and to budget, often against a background of divergent and conflicting objectives.  As a successful outcome depends on team members accepting direction and giving up some of their autonomy, the development process needs to foster team members' abilities to develop shared commitments and to work collectively to achieve them.

Ideally, the process converges efficiently towards good solutions which everyone (including the client) understands and accepts. In practice, there are plenty of opportunities for things to go wrong, e.g. through a failure to understand the constraints in the domain; lack of knowledge about other subsystems; continual fluctuations in requirements; poor planning; mismanagement of the development process, and so on. Any approach to human error in safety-critical software development needs to encompass the variety of sources of errors which exist in such an intellectual, social and organisational process.

The following section reviews the current approaches to human error and discusses their limitations. Section Three discusses ways of viewing human error which the PRICES project has found to be more relevant for intellectual teamwork and which have been taken into account in developing the PRICES Code of Practice.


# 2 Current Approaches to Human Error

## 2.1 Human Error as Unreliable Behaviour

The Human Reliability Assessment (HRA) community has established an impressive track record of work in human error, which it views as the unreliable behaviour of a fallible machine (Hollnagel, 1993). Hollnagel (1993) provides a good overview and comparative evaluation of various HRA methods. The goal of HRA is to identify all points in a sequence of operations where incorrect (or failed) human action may lead to adverse consequences (Swain, 1990). Probabilities are assigned to each event, and aggregation of these yields an overall figure for the probability of human error over the whole chain of events. It is usually a three-stage process:
1. Task Analysis (often Hierarchical Task Analysis).
2. Human Error Analysis from Risk Assessment field e.g. HAZOP[3].
3. Human Reliability Quantification - estimating error probabilities from historical data e.g. of incidents or using HRA techniques such as HEART[4] or THERP[5] based on generic error probabilities.

HRA practitioners take account of the work context in the guise of Performance Shaping Factors (PSFs). PSFs are the (internal and external) factors, e.g. working procedures, which affect the way that the individual executes the task, making errors more or less likely (Swain, 1988).

HRA techniques are probabilistic and behaviouristic (Mancini, 1990), hinging on the belief that probabilities attached to subtasks in a sequence of

---

[3] HAZOP: Hazard and Operability Study.

[4] HEART: Human Error Assessment and Reduction Technique.

[5] THERP: Technique for Human Error Rate Prediction.

operations can be aggregated. It has been argued that human actions are not additive and independent, but are more likely to be conditional on some basic assumptions being true (Hollnagel, 1993). The probabilities of certain actions occurring are not necessarily stable but may vary enormously with factors such as time of day, morale etc. (Moray, 1990).

Despite these limitations, the techniques have been used with genuine success for analysing manual, repetitive tasks. However, when reviewing HRA techniques for use in the Nuclear Industry, the ACSNI Study Group (1991) concluded that they were less useful for analysing intellectual tasks:

"It will be evident that HRA is still in its infancy... Assessment will need to be supplemented by awareness of the possibility of *false decision, rather than failure of skill.*" (ACSNI Study Report, 1991 p.viii). (italics added)

## 2.2 Human Error as Mishaps in Problem-Solving

The human error model most often cited by HRA practitioners is the one developed by Rasmussen (1969, 1974, 1983). Rasmussen developed his three-level model by analysing the errors made by seasoned operators on the basis of an extensive review of incident and accident reports from hazardous industries. He categorised the errors according to whether or not: procedures were followed; necessary information was available in an appropriate form; and, information was wrongly interpreted (Sanderson & Harwood. 1988).

Rasmussen was heavily influenced in his thinking by the (then) fashionable early artificial intelligence (AI) work on thinking as problem-solving. He concluded that an individual would use a skill to deal with a problem-free task; use rule-based behaviour for handling a routine problem, and resort to first principles to deal with a novel problem. This is shown in Table 1.

| Degree of Problem | Type of Behaviour |
| --- | --- |
| A novel problem is encountered | Knowledge-based i.e. use of first principles |
| A routine problem is encountered | Rule-based |
| No problem encountered | Skill-based |

Table 1: Showing Rasmussen's (1983) Skills-Rules-Knowledge Model

The model has a number of limitations. Firstly, determining a situation to be 'problematic' is an individual judgement, not easily extrapolated from one person to another. Secondly, the 'skill' category appears to be hiding a whole host of abilities - ranging from routinized, automatic skills e.g. scanning, to acts of creativity emanating from outside of conscious awareness.   Thirdly, it is unclear whether these levels are ever-present levels associated with different types of mental functioning or stages of learning. Rasmussen has argued it both ways by crediting his model with a learning element, but all of the operators in Rasmussen's study were fully trained, so a learning perspective is hard to maintain.

Lastly, the model is weak in problem-formulation.   AI scientists eventually realised that the difficult part of problem-solving does not lie in the mechanical part of the process (listing alternatives, choosing between them, etc.) but in formulating the problem in the first place.   During the 1980s, Rasmussen's basic model underwent adaptation by cognitive error psychologists such as Reason (1990).  It is Reason's version, described next, which is most often cited by the safety-critical software engineering community.


## 2.3  Human Error as Dysfunctional Planning

Reason (1990) argued that human errors should not be defined superficially, but in terms of their underlying cognitive mechanisms e.g. retrieval failures.   His major contribution to this field is that he linked Rasmussen's SRK Model to error types and associated cognitive mechanisms (Dougherty, 1990).

Reason developed a model whereby basic error tendencies e.g. resource limitations, interact with information processing domains, such as recognition, to produce primary error groupings e.g. inaccurate recall. He proposed five basic error tendencies and eight primary error groups, and abstracted the eight primary error groupings into two types - mistakes and slips or lapses. Mistakes result from a wrong decision about what action to undertake, whilst slips or lapses result from a correct decision wrongly executed.   Mistakes were subdivided into:  (a) *failures of expertise*, where some pre-established plan or problem solution is applied inappropriately and (b) a *lack of expertise*, where the individual, not having an appropriate 'off-the-shelf' routine, is forced to work out a plan of action from first principles.  These two correspond to the rule-based and knowledge-based levels of performance as described by Rasmussen (1983), shown in Table 2 on the next page.

In Reason's model, human error is bound up with the notion of intention, as Reason addressed only planned actions that fail to achieve their desired consequences without the intervention of some chance or unforeseeable agency. Reason characterised the decision-making process as having four parts: (1) the setting of objectives; (2) the search for alternative courses of action; (3) comparison and evaluation of alternatives;  and (4) decisions about appropriate action.

| Nature of Problem | Cognitive Basis of Behaviour | Type of Errors associated with each behaviour | In terms of Expertise | Locus of Cognitive Mechanism Failure |
|---|---|---|---|---|
| Novel - never seen before | Resorts to first principles based on existing <u>knowledge</u> | Mistakes | Lack of expertise | Planning |
| Routine or familiar | Uses <u>rules</u> which have worked before | Mistakes | Misapplied expertise | Planning |
| None | Uses an existing <u>skill</u> | Slips and lapses | Lapse of expertise | Execution and/or storage |

**Table 2:  Showing Reason's Model in relation to SRK and to Expertise**

The model has several limitations.  Firstly, it is based on the Hamlet model of decision-making with an unusual amount of time available for deliberation and planning (Dreyfus, 1985).  This is known to be inappropriate for most organisational decision-making (Eccles et al, 1992). Whilst managers may agree on the value of systematic evaluation of choices, the reality is that most of their decisions have to be taken in shorter timespans with unclear consequences.  Note that this criticism of the view of decision-making as always involving systematic planning does not extend to the usefulness of plans as resources for action e.g. project plans.

Secondly, the model doesn't allow for those instances where misunderstandings arise between individuals - arguably the most common form of error in a working environment.  Thirdly, the emphasis on cognitive mechanisms may not be applicable to intellectual work, where the quality of the ideas which are generated play a vital role in how the work is conceptualised and carried out, and any resource limitations that do exist may not necessarily be fixed.  The intellectual skills normal people display are limited, in practice, not so much by innate restrictions to their various 'channel capacities', as by the amount of motivating energy they are able to bring to bear on the task in question (Hudson, 1975).

Reason's model offers 'explanations' for the occurrence of errors in terms of mechanical breakdown of the mind's cognitive mechanisms, which (whilst of undoubted interest to cognitive psychologists) would not satisfy accident investigators who are far more interested in attributing responsibility and negligence.  The model does usefully distinguish between mistakes made because of a lack of ability and lapse-of-concentration errors.  Software errors resulting from attentional deficits during the development process are usually less significant than mistakes as they tend to be spotted during modular testing or review and are rarely found to have any safety implication.

All of the approaches outlined in this section are interrelated, having an overlapping history of development. They focus either on repetitive manual tasks undertaken by individuals or they 'explain' errors in terms of cognitive failure. As such, they are of only marginal interest for intellectual teamwork such as software engineering. Even where the model developers have moved on e.g. Reason is now more interested in violations of working procedures and organisational learning, the software engineering community persists in citing the human error models as a basis for its approach. This may be because it is not clear how to merge these new wider interests with the original models.

# 3  Other Perspectives on Human Error

This section examines alternative views of human error which have been neglected by error psychologists, but which the PRICES project has drawn from to underpin the Code of Practice.

## 3.1 Human Error as Misjudgement

"Our reacting to the present bombardment of information involves ignoring some of it, seizing the rest and interpreting it in the light of past experience in order to make as good a guess as possible about what is going to happen. This may be called a process of *judgement*: that is, making a decision or conclusion on the basis of indications and probabilities when the facts are not clearly ascertained." (Abercrombie, 1989 p.14)

Reason (1990) defined expertise as an ability to plan effective action. This planning model of decision-making dominates cognitive psychology, but has been criticised by the Dreyfuses (1985). They contend that deliberated planning of decisions is not a feature of fully developed expertise, but is characteristic of an earlier stage of learning. The Dreyfuses (1985) developed their own model of expertise charting the evolution of promising novices into experts. It is based on their findings from studying developing expertise in a range of professional types, and is in line with the findings of others e.g. Abercrombie (1989). An outline of the model is shown in Table 3 on the next page. Different types of errors can be associated with each stage of development.

The model disputes what current models purport i.e. that learning a skill or developing expertise involves moving from [1] deployment of knowledge of first principles, to [2] rule-based behaviour, to [3] skill-based behaviour. Instead, this adult-learning model depicts a movement from [1] rule-based behaviour to [2] competent planning to [3] intuitive involvement in strategic choices, where the hallmark of expertise is the ability to choose an appropriate strategy with a successful outcome.

| Stage | Characteristics of each Stage of developing Expertise |
|---|---|
| 1. Novice | The novice learns to recognise various objective facts and features relevant to the skill and acquires rules for determining actions based upon those facts and features.  Relevant elements are so clearly and objectively defined for the novice that they can be recognised without reference to the overall situation in which they occur.  Such elements are 'context-free', and the rules applied to them 'context-free rules'. |
| 2. Advanced Beginner | Thanks to a perceived similarity with prior examples, the advanced beginner starts to recognise the presence (or absence) of certain 'situational' elements.  The advanced beginner can distinguish the salient features of the situation, but has little feel for what is important. |
| 3. Competence | The competent individual copes with the bombardment of a huge array of information by adopting a hierarchical procedure of decision-making i.e. by choosing a plan to organise the situation, and by then examining only the small set of factors that are most important given the chosen plan.  This is the Hamlet model of decision-making - the detached, deliberate, and sometimes agonising selection among alternatives, which bears no resemblance to the higher levels of expertise characterised by a rapid, fluid, involved kind of behaviour. |
| 4. Proficiency | Here, the individual approaches the task from some specific perspective gained through experience.  As events modify the salient features - plans, expectations, and even the relative salience of features gradually change.  No detached choice or deliberation occurs - it develops seemingly naturally.  This stage is marked by a combination of intuitive understanding, and analytical thinking about the situation. |
| 5. Expertise | Whilst most expert performance is ongoing and nonreflective, when time permits and outcomes are crucial, an expert will deliberate (critically reflect on intuitions) before acting.  Even after critical reflection, an expert's decision may not work out, as (s)he may be thrown by a curve of events that could not have been foreseen. |

**Table 3:  Showing an outline of the Dreyfuses (1985) Model of Expertise**


Finally, expertise has a strong social component.  Abercrombie (1989) observed that trainees developed understanding of their craft by reflecting on disparities between their own results (and on how the results were achieved) with those of others who were more skilled.  Professional relationships provide a context for the formation, testing and modifying of ideas, ensuring that the expertise that develops is appropriate for the field of endeavour. It affords the individual some (but not complete) protection from undue pressures that others may exert, but one's own outlook (developed from experience) continues to exert unnoticed influence.

## 3.2 Human Error as Bias

"We are doomed to receive far more information than we could [sic] cope with, and therefore need to select, and in selecting, any yardstick is better than none. Our capacity for intuitive, over-simplified judgement is basic to our survival." (Hudson, 1975 p.68)

Such 'over-simplified judgements' are biases, which can be endogenous or exogenous. Endogenous biases accrue from personal experience - people are inclined to perceive the world through the lens of past events creating self-fulfilling prophecies (Tversky & Kahneman, 1974), and may experience a strong motivating force to reduce the cognitive dissonance which ensues from receipt of information that conflicts with existing meaning structures (Festinger, 1957). Festinger's theory has been updated by Billig (1989) to take account of its (neglected) interpersonal aspect, arguing that dissonance is created by a need to justify choices to others, rather than to oneself. This is representative of the second (exogenous) source of bias, emanating from the expectations and influence of others.

Inaccurate meaning structures can sustain unreasonable conviction despite inadequate or contradictory evidence (Ayton & Pascoe, 1995; Plous, 1993). This conviction can be evidenced in a variety of tendencies, such as:
- failing to seek evidence which could disconfirm an initial opinion. This may result in self-fulfilling prophecies (Merton, 1948) or Pygmalion effects - misconceptions which ultimately prove true (Rosenthal & Jacobsson, 1968);
- failing to appreciate the negative consequences of a chosen path;
- misperceiving that too much has been invested to change direction;
- prejudicial focusing of attention resulting in misattributing causality;
- extrapolating from small samples or single instances;
- attributing undue importance to recent events i.e. recency effects.

Hosking and Morley (1991) discuss some of the moves that people make to preserve stable, simple and consistent cognitions. These include:
- engaging in various forms of wishful thinking;
- relying on a small number of highly salient cues;
- seeking out evidence to strengthen existing beliefs;
- overweighting evidence which is consistent with beliefs;
- underweighting evidence which is not; and
- polarizing judgements.

Effective debiasing strategies usually entail consideration of alternative perspectives, leading to more rounded decisions (Plous, 1993). Research on social facilitation (Zajonc, 1965; Henchy & Glass, 1968) and diffusion of responsibility tends to support Tetlock's (1985) notion of organisational decision makers as finely tuned politicians. In brief, performance is sensitive to many social factors, which may enhance or hinder it.

When groups are cohesive and relatively insulated from the influence of outsiders, group loyalty and pressures to conform can lead to 'Groupthink': Deterioration of mental efficiency, reality testing, and moral judgement that results from in-group pressures (Janis, 1982). Janis advocated changing the mode of operation, but others have proposed tinkering with the group composition e.g. Belbin (1981). Belbin's theory has received populist support, but is most applicable to teams which are more unstructured than is usual in software development where roles and responsibilities are often clearly defined. Belbin's study (and the work of others e.g. Steiner, 1972) does show that the accuracy of team judgements depends on a variety of factors, including the nature and difficulty of the task, the competence of group members, the activity of the leader, and the way that group members are allowed to interact etc. (Plous, 1993).

## 3.3  Human Error as Misunderstanding

Misunderstandings are to be expected on any safety-critical software development project, but are especially prevalent on those which are large-scale, require expertise from several different disciplines, and which have staff distributed over a wide geographical area. Such large-scale, safety-critical, embedded software development projects are multi-disciplinary, at the intersection of software engineering; systems engineering and the application domain. Each of these disciplines has its own vocabulary, working practices, assumptions, perspectives, values, models and tools. A common language has to be found before a common understanding can be forged. Adequate working relationships cannot be established without this understanding, but in the meantime, time pressures require that decisions be taken about what the system and subsystems should do, and how the work should be organised and meshed together.

It has been suggested that most software errors with safety-critical implications arise from misunderstandings between individuals and misunderstandings between teams about what the software should do (Lutz, 1992)[6]. When discovered, these misunderstandings about what is required often necessitate expensive rework. A reduction in this type of safety-critical software error is dependent on such misunderstandings being minimised, if not eliminated.

---

[6] Technically, Lutz looked at mission-critical (unmanned spacecraft) rather than safety-critical projects.

Anthropologists define culture as: A system of shared meanings and shared understandings (Moerman & Sacks, 1971), and would suspect the gaps between (sub)cultures of being the breeding ground for misunderstandings. As organisational culture and structure are known to be intertwined, reducing misunderstandings is dependent on creating an appropriate structure which removes or mitigates any organisational or geographic barriers. This new structure should facilitate better communication - not just of information flow but also enabling closer working relationships to be forged across traditional divides. Whilst superficial levels of culture may be changed by mandate, changes at a deeper level without corresponding structural alterations are rarely successful.

There has been a recent recognition that the prevailing organisational culture should embody a safety element (Leveson, 1995). Safety culture can be viewed as the assembly of characteristics and attitudes in organisations which establishes that, as an overriding priority, safety issues receive the attention warranted by their significance. There are a number of identifying signs or marks of a safety culture, which include a management safety policy, management of risk, and fault prevention policies and procedures (Turner, Pidgeon, Blockley & Toft, 1989). Fletcher et al (1991) have produced an audit checklist for this purpose. The creation of a safety culture cannot be achieved without sincere management commitment as employees will quickly sense where management's true priorities lie and may conform to these even when they conflict with explicit policy statements.

## 3.4 Human Error as Mismanagement

"In most of the major accidents of the past 25 years, technical information on how to prevent the accident was known and often even implemented. But in each case, the technical information and solutions were negated by organizational or managerial flaws." (Leveson, 1995 p.48)

Most disaster investigations have found that, in the final analysis, the cause of the disaster could not be attributed to an error made by a single individual, but arose from more systemic organisational or managerial flaws (Pidgeon, 1988; Pidgeon and Turner, 1986). Management culpability can take (at least) two forms.

Firstly, it can arise from a failure of strategic planning, implementation of plans or the inappropriate deployment of resources. This may be coupled with an inability to hear unpalatable messages about technical difficulties when hemmed in by budgetary constraints and an imperative to proceed.

Secondly, it can arise as a consequence of failing to provide a working environment conducive to error-free work or by not establishing appropriate working practices, e.g. failure to adhere to Standards such as IEC 1508, ISO 9000.

Whilst management culpability appears to be an established feature in accident scenarios, it is still missing from human error models. Within the industrial sphere it has been necessary to widen the term to include management error (Embrey et al, 1994). The PRICES project consider it to be necessary to also widen the term when referring to errors made in the software development process, emphasising that errors can emanate from diverse points in the system. Inappropriate organisational procedures and insensitive management can have devastating effects on safety and often underlie individuals' errors.

## 4    Summary of PRICES Approach

PRICES found the traditional approaches to human error to be too individualistic to underpin the Code of Practice. Drawing from the research cited above makes it possible to consider, in a theoretically coherent way, sources of potential errors within teams, between teams, in inappropriate organisational culture, structure and managerial deficits. Further, at the individual level, the Dreyfuses (1985) model of expertise is an attractive alternative to traditional approaches to error, providing as it does, a convincing account of how people learn to reason about complex intellectual tasks.

As little can be learnt without some degree of 'trial and error', errors are a natural occurrence in this model and have no need of explanation in terms of faulty cognitive mechanisms. Instead, the (hazardous) conditions e.g. task-ability mismatch, which encourage people to reason one way or another (for better or worse) need to be analysed, understood and managed. This perspective is somewhat analogous to that which is now recommended by safety engineers such as Kletz (1991) who advocates managing error in industrial environments by changing the working conditions and not by trying to shoehorn people into ill-fitting environments.

The developers of the PRICES Code of Practice are of the opinion that the identification and control of these hazardous conditions constitutes human hazard management rather than human reliability assessment. In this, PRICES has taken a lead from safety engineers (Comerford & Blockley, 1992) who argue that the notion of assessment does not necessarily imply any feedback loop or ongoing vigilance, whilst hazard management does. In addition, the term reliability is more applicable to manual tasks than to intellectual ones. A human hazard management approach has been taken in developing the Code of Practice, where some, though not all, of the hazards are the potential for inserting safety-critical errors in the software product.

# 5    Conclusion

Safety-critical software development is an intellectual, social and organisational process which can give rise to a diversity of human errors. All source of errors (and not just those made by individual software engineers) need to be considered in an integrated way. PRICES has taken account of this diversity in deciding what type of guidance needs to be provided in the Code of Practice. The advice itself has not been derived theoretically but represents the results of practical work undertaken on PRICES and other projects in the Safety-Critical Systems Research Programme.

# References


Abercrombie M.L.  The anatomy of judgement:  An investigation into the processes of perception and reasoning.  London:  Free Association Books, 1989.

ACSNI  Study group on human factors, second report:  Human reliability assessment - a critical overview.  Advisory Committee on the Safety of Nuclear Installations, 1991.

Ayton P. and Pascoe E.  Bias in human judgement under uncertainty.  The Knowledge Engineering Review 1995; Vol.10, 1, pp21-41.

Belbin R.M.  Management teams:  Why they succeed or fail.  London: Heinemann, 1981.

Billig M.  Arguing and thinking:  A rhetorical approach to social psychology. Cambridge: Cambridge University Press, 1989.

Comerford J.B. and Blockley D.I.  Managing safety and hazard through dependability.  Internal report, Department of Civil Engineering, Bristol University, 1992.

Dougherty E.M. Jr.  Journal of Reliability Engineering and System Safety. Special Issue on Human Reliability Analysis 1990; Vol.29, pp283-300.

Dreyfus H.L. and Dreyfus S.E.  Mind over machine.  Free Press:  New York, 1985.

Eccles R.G., Nohria N. with Berkley J.D.  Beyond the hype:  Rediscovering the essence of management.  Harvard Business School Press, 1992.

Embrey D. E.  Kontogiannis T. & Green M.  Guidelines for preventing human error in process safety.  Centre for Chemical process Safety, American Institute of Chemical Engineers, New York, 1994.

Festinger L.   A theory of cognitive dissonance.   Stanford:   Stanford University Press, 1957.

Fletcher B., Jones F. and Turner G.   Measuring organisation culture:  The cultural audit.   Internal report, Department of Psychology, University of Hertfordshire, 1991.

Henchy T. and Glass D.C.   Evaluation apprehension and the social facilitation of dominant and subordinate responses.  Journal of Personality and Social Psychology 1968, 10, pp446-454.

Hollnagel E.   Human reliability analysis:   Context and control.   UK: Academic Press, 1993.

Hosking D. M. and Morley I. E   A social psychology of organizing.  UK: Harvester Wheatsheaf, 1991.

Hudson L.  Human beings:  An introduction to the psychology of human experience.  UK: Triad Paladin, 1975.

Janis I.L.  Groupthink:  Psychological studies of foreign policy decisions and fiascoes.  Boston:  Houghton-Mifflin, 1982.

Kletz T.   An engineer's view of human error.   Institution of Chemical Engineers.  VCH Publishers: NY, 1991.

Leveson N.G.  Safeware:  System safety and computers.  Addison-Wesley, USA, 1995.

Lutz R.   Analyzing software requirements errors in safety-critical, embedded systems.  Internal Report, Jet Propulsion Laboratory, California Institute of Technology, Pasadena, CA 91109, USA 1992.

Mancini G.   A solution to Dougherty's apprehensions?   Journal of Reliability Engineering and System Safety.   Special Issue on Human Reliability Analysis 1990 Vol.29, pp329-336.

Merton R.K.  The self-fulfilling prophecy.  Antioch Review 1948, 8, pp193-210.

Moerman M. and Sacks H.  On understanding in conversation.  Presented at 70th Annual Meeting, American Anthropological Association, NY City, 1971.

Moray N. Dougherty's dilemma and the one-sidedness of human reliability analysis.  Journal of Reliability Engineering and System Safety.  Special Issue on Human Reliability Analysis 1990, Vol.29, p337-344.  UK:  Elsevier Science.

Pidgeon N.  Risk assessment and accident analysis.  Acta Psychologica 1988, 68, pp355-368.

Pidgeon N. and Turner B.A.  "Human error" and socio-technical system failure.  In: A.S. Nowak (ed.) Modeling human error in structural design and construction.  NY: American Society of Civil Engineers, 1986.

Plous S.  The psychology of judgment and decision making.  McGraw-Hill: USA, 1993.

Rasmussen J.  The human data processor as a system component. Bits and pieces of a model, Riso-M-1722.  Riso National Laboratory, Roskilde, Denmark, 1974.

Rasmussen J.  Man-machine communications in the light of accident reports.  Riso Reports S-1-69.  Riso National Laboratory, Roskilde, Denmark, 1969.

Rasmussen J.  Skills, rules, knowledge: signals, signs and symbols and other distinctions in human performance models.  IEEE Transactions: Systems, Man & Cybernetics 1983, SMC-13, pp257-267.

Reason J.  Human error. Cambridge University Press, 1990.

Rosenthal R. and Jacobson L.  Pygmalion in the classroom: Teacher expectations and pupils' intellectual development.  New York: Holt, Rinehart & Winston, 1968.

Sanderson P.M. and Harwood K.  The skills, rules and knowledge classification: A discussion of its emergence and nature.  In: Goodstein L.P., Andersen H.B. and Olsen S.E. (eds.) Tasks, Errors and Mental Models. London: Taylor & Francis, 1988.

Swain A.D.  Adapting risk analysis to the needs of risk management. Paper presented at the World Bank Workshop of Risk Management and Safety Control, Washington, DC, 1988.

Swain A.D.  Letters to the Editor.  Journal of Reliability Engineering and System Safety.  Special Issue on Human Reliability Analysis 1990, Vol.29, pp371-374. UK: Elsevier Science.

Steiner I.D.  Group process and productivity.  New York: Academic Press, 1972.

Tetlock P.E.  Accountability: The neglected social context of judgment and choice.  Research in Organizational Behavior 1985, 7, pp297-332.

Turner B.A., Pidgeon N., Blockley D.I. and Toft B.  Safety culture: Its importance in future risk management.  Position paper for 2nd World Bank Workshop on Safety Control and Risk Management, Karlstad, Sweden, 1989.

Tversky A. & Kahneman D.  Judgment under uncertainty: Heuristics and biases.  Science 1974, 185, pp1124-1130.

Zajonc R.B.  Social facilitation.  Science 1965, 149, pp269-274.

# The StAR Risk Adviser: Psychological Arguments for Qualitative Risk Assessment

Peter Ayton and David K. Hardman
Department of Psychology, City University
London, United Kingdom

## Abstract

Given reliable statistical evidence, calculations of quantitative values of risk may be attempted. Novel technologies however are, by their nature, characterised by a lack of appropriate historical data which creates difficulties for conventional quantitative approaches to risk assessment. Numerical risk assessments which convey spurious validity and precision can be dangerously misleading. Here we describe and explain the psychological rationale for a decision support system called StAR that utilises quantitative information where appropriate, but which is also able to use qualitative information in the form of arguments for and against the presence of risk. The user is presented with a summary statement of risk, together with the arguments that underlie this assessment. Furthermore, the user is able to search beyond these top-level arguments in order to discover more about the available evidence. We argue that this approach is well-suited to the way in which people naturally reason about decisions, and we show how the StAR approach has been implemented in the domain of toxicological risk assessment.

## 1. Introduction: assessing technological risk

Although the advance of science and technology in developed nations has brought with it better health and longevity, it also continues to present people with new hazards to fear. The possibility that one will be exposed to such a hazard is a type of *risk*. Although a degree of risk-taking is often desirable or even essential to the development of human society [Zeckhauser and Viscusi 90], it is also of increasing concern to governments, organisations, and the general public that the risk from new technologies be minimised as far as is practicable [Royal Society 92; Health and Safety Executive, 96]. Only by conducting a proper risk assessment is it possible to decide how to minimise risks affecting the public, whether this be by finding methods for managing the risk or by abandoning the technology.

Many of the risks we face are easily quantified because there is reliable statistical evidence on which to base a judgement. For example, the probability of being involved in a motor vehicle accident can be estimated from the most recent statistics. Where a technology is new, however, it is not always so easy to estimate risk simply because there is no historical data. Sometimes statistical models can overcome this problem to a large extent; for example, in the case of electromechanical systems, precise and reliable estimates of the likelihood of system failure can be derived from historical data on the failure rates of standard components or from simulations of system behaviour.

For some technologies, however, current methods for risk assessment are less satisfactory. Consider the case of chemical compounds used in foods and agriculture (an example we shall be returning to throughout this paper). Formulation of policy on the basis of quantitative models is highly problematical because there is often sparse knowledge of the nature and biological action of complex chemicals. The Department of Health's Committee on Carcinogenicity of Chemicals in Food [Carter 91] concludes:

> "The committee does not support the routine use of quantitative risk assessment for chemical carcinogens. This is because the present models are not validated, are often based on incomplete or inappropriate data, are derived more from mathematical assumptions than from a knowledge of biological mechanisms and, at least at present, demonstrate a disturbingly wide variation in risk estimates depending on the model adopted."

In everyday life, experts are sometimes unable (or, perhaps more accurately, unwilling) to provide probabilities to describe risks, as the following example illustrates. On the 26th of March 1996, following the revelation that the consumption of BSE-infected[1] beef was the likeliest cause of a new strain of Creuzfeldt-Jacob Disease[2], the UK government's Spongiform Encephalopathy Advisory Committee issued a statement to British newspapers concerning the risks associated with British beef. This statement explicitly avoided mention of any quantification of the risk of BSE infecting beef-eating humans: "The committee has carefully considered whether a quantitative risk assessment can provide an estimate of the absolute risk in relation to BSE. In its judgement a precise measure is impossible because of a number of interacting uncertainties". However, the same report also stated that "The risk is likely to be extremely small".

Concerns over the application of statistical and other quantitative methods are not confined to the field of chemical carcinogenicity; many professionals, managers and policy makers worry that the assumptions behind such techniques are often

---

[1] BSE refers to Bovine Spongiform Encephalopathy, a degenerative brain disease in cattle, popularly known as "mad cow disease".

[2] Creuzfeldt-Jacob Disease (CJD) is a degenerative brain disease in humans, often referred to as the human equivalent of mad cow disease. At the time of writing there is a strong (but unproven) possibility that contact with BSE-infected material can cause certain types of CJD.

simplistic and offer an illusion of precision[3]. Moreover, even where quantitative techniques are not disputed, policy makers and the public may find it difficult to understand the mathematical justifications behind risk assessments, as well as their implications, due to the technical language used to convey risk information. As *Newsweek* (May 12th, 1986) said of radiation risk statistics: "The numbers don't really mean anything beyond that disaster could happen at any time". If it were possible to convey risk information in a manner that is easily comprehensible to people, then the understanding of the nature of the risk might be improved.

Where considerable uncertainty exists "informed judgement" may be a better way of formulating decisions, plans, and policies. And indeed, several management technologies have been developed which aim to elicit and exploit *judgmental* numerical information. For example, decision analysis [Raiffa 68], cross-impact analysis [Dalkey 72] and fault-tree analysis [Fischhoff, Slovic and Lichtenstein 78] all attempt to provide quantitative conclusions from judgmental inputs. *Subjective* probabilities are one of the prime inputs into these and many other management technologies. However there are grounds for questioning the viability of this approach as a panacea for risk assessment. A large literature attests to the difficulties that people - including experts - have in producing valid quantitative assessments of likelihood. These difficulties are known to be particularly acute when people are required to assess a likelihood for a proposition or unique event rather than a relative frequency [cf. Ayton and Pascoe 95].

In many risk assessments there are events of interest for which no obvious reference class exists and then one will plainly be unable to assess a statistical likelihood according to the frequentist definition of probability. Consider, for example, the possibility, in 1991, that Saddam Hussein would attack Kuwait. How could President Bush's administration have gone about assessing a subjective probability for this unique proposition? There is no obvious class of events to which this event belongs. As van der Heijden [94] discusses, such an assessment task may place unrealistic demands on the forecaster. van der Heijden argues that in planning for such plausible, high consequence, and unique events, quantitative approaches are intuitively less relevant because natural thinking about the future doesn't involve consideration of probabilities but entails reasoning by extrapolating models of the world. As a consequence van der Heijden advocates the application of scenario planning techniques which aid the creation of a robust strategy that works well under a range of plausible futures. As a methodology for dealing with uncertainty, scenario planning accepts and downplays the decision maker's poor ability to make realistic probability assessments for single events. Ecologically, the acceptability of scenario planning techniques to senior managers, and the relative disdain with which quantitative decision analysis is viewed, may reflect an intuitive appreciation of the poor quality and limited utility of the quantitative approach.

---

[3] In recounting the investigation into the *Challenger* space shuttle disaster, Richard Feynman [Feynman 88] reports that the risk estimates of NASA management and engineers differed by a factor of more than 300. The more remote estimate provided by the management led Feynman to comment: "If a guy tells me the probability of failure is 1 in $10^5$, I know he's full of crap" (Page 216).

Over recent years designers of expert systems, as well as management scientists, have been developing tools and methods to aid, or even replace, human decision making. Although both communities are focused on similar problems, the tools and general frameworks for analysis tend to differ. Perhaps because of the rather separate disciplinary backgrounds of management scientists and expert systems designers, two rather distinct and independent approaches have emerged. The decision analytic approach emphasised by management science relies heavily on quantitative estimation of values and calculation using normative decision rules, while expert systems have utilised predominantly qualitative representations of particular knowledge about causality, time and constraints. In the next section we review the psychological evidence which indicates something of the qualitative nature of natural human thought when applied to problems involving uncertainty. We believe that this work has important implications for the design of decision support systems in general. We argue it gives a rationale for the qualitative approach to risk assessment that is adopted by StAR.

## 2.  Reasoning under uncertainty

### 2.1  Normative, descriptive, and prescriptive approaches

*2.1.1. The normative approach*

In the psychology of judgement and decision making it is common to distinguish three approaches: the normative, the descriptive, and the prescriptive. The normative approach is the study of guidelines for right action. It involves the search for, identification of and defence of principles usually expressed as rules that individuals ought to follow. With such rules one can specify a method for reaching an optimum solution to a problem. For example, rational choice theory [e.g. Edwards 54] specifies that decisions under uncertainty should proceed by maximising *subjective expected utility* (SEU). In other words, any decision involving probabilistic outcomes is best calculated by (1) multiplying the personal value of each outcome by how likely you think that outcome is to occur, (2) choosing the course of action for which the procedure specified in (1) gives the highest value. However, there is considerable evidence that the SEU approach does not capture the way in which people actually make uncertain choices or judgments. At a recent conference involving the leading decision researchers Edwards [92] took two straw votes and found that the conference unanimously endorsed SEU as the appropriate normative model and also unanimously agreed that people don't act as the model requires.

If people don't behave as the normative theories imply they should then there is clearly a need for an account of what it is that people are doing and why they deviate from the normative ideal. In fact a descriptive account would be needed regardless of the nature of human judgement. Normative theories are no more accounts of *how* people make decisions than the rules of chess are explanations of how people decide on a strategy to beat an opponent. We turn to consider the descriptive approach to decision making which is the study of the decisions that people actually make and how they decide.

## 2.1.2. The descriptive approach

So why don't people act in accordance with what the normative model requires? In the first place, it seems that people do not necessarily have a stable ordering for preferences between options [e.g. Tversky and Simonson 93; Tversky and Shafir 92a] and, in fact, it seems that preferences are often constructed at the time judgements are made, rather than simply computed from stable values held in memory, in response to a judgement or choice [Slovic, Griffin, and Tversky 90; Payne, Bettman, Coupey, and Johnson 92]. In a comparison of choice and bidding procedures Lichtenstein, Slovic and Zinc [69], Lichtenstein and Slovic [71; 73] and Tversky, Sattath and Slovic [88] offered subjects choices between gambles like the following:

*P bet*            *29/36 probability to win $2*

*$ bet*            *7/36 probability to win $9*

These authors find that subjects choose the P bet but are willing to pay more to obtain the $ bet. Subjects prefer the P bet because of its odds but then set a higher value on the $ bet because of its larger winning payoff. This is described by Slovic [72] as a "compatibility effect"; since a monetary value is expressed in terms of money, subjects find it easier to use the monetary aspect of the gamble to set the value of the gamble. When subjects are asked which they prefer they have no particular reason to use the monetary aspect of the gamble to set the value of the gamble. A similar phenomenon labelled the Prominence effect by Tversky et al [88] illustrates the potential significance of this decision strategy. Tversky et al found that the weight given to predominant attributes (those features considered the most important by the subjects) increased when preference is assessed with choice rather than a matching task.

Subjects were asked to imagine they had to decide between two options proposed for reducing traffic accidents. Both options were described in terms of yearly costs and the number of casualties that the current (unspecified) high level would be reduced to per year.

<u>Choice task</u>

*Option A:*      *570 casualties*          *cost $12 million*

*Option B:*      *500 casualties*          *cost $ 55 million*

Most people preferred B to A. This pattern of preference implies that the difference in casualties (70) is more important than the difference in costs ($43 million). However, changing the method of judgement to a matching task significantly changed the judgments. Thus when a different group of subjects were asked to consider how much money they would be prepared to pay for a scheme that reduced accidents to 500 casualties such that they were indifferent between that scheme and Option A (570 casualties; cost $12 million) they typically gave a value that was far below $55 million.

<u>Matching task</u>

*Option A:   570 casualties   cost $12 million*

*Option B:   500 casualties    cost $ x*

The trade-off between attributes was different with matching than with choice. Why did this happen? Tversky et al argue that the choice task invites more qualitative reasoning - so people select the option that is superior on the most important attribute. This is cognitively easier, easier to justify and conflict avoiding. But matching requires a more quantitative assessment. In order to perform the matching task at all attention *must* be paid to the values of both attributes and their relative importance. The general point to be drawn from this research is that the psychological context of judgments, and the mode by which they are assessed, can have critical effects on the judgments themselves. The reason for this is that the natural mode of reasoning about such problems does not involve computation by the manipulating of static values for utility but instead entails a more qualitative reasoning process.

A second reason for human decisions deviating from the normative ideal is that subjective probabilities are, by their very nature, not objective! In some circumstances this may not be of much importance, but when we are evaluating possibly serious risks that affect society then we need to know that a probability judgement is as accurate as possible. Unfortunately, there is evidence that some kinds of probability judgments can be biased, and that the person making the judgement may be overconfident as to the accuracy of that judgement [for reviews see: Ayton and Pascoe 95; Bolger and Wright 94; McClelland and Bolger 94]. Where people are asked to estimate the likelihood of a single event it seems that they may not reason in accordance with any valid statistical principles. For example, Tversky and Kahneman [83] asked professional statisticians to rate the probabilities of various events, some of which were inclusive of others. Thus, the probability that next year there will be an earthquake in San Francisco causing a dam burst resulting in a flood in which at least 1000 people drown *must* be less than the probability that there will, next year, be a flood *somewhere* in the USA [including San Francisco] in which at least 1000 people drown. However, the professional statisticians typically rated the logically less likely event as more likely than the more general event set of which it is a member. It seems that the provision of a plausible causal scenario in the former case (i.e. an earthquake in San Francisco) is a salient consideration in the judgments that subjects make. The nature of the descriptions of events affect how easy it is to imagine them and ease of imagination affects estimates of likelihood for single events [Tversky and Kahneman 83].

Descriptive theories of human judgement and decision making attempt to capture the nature of what it is that people actually do. For example, Kahneman and Tversky [79; 81] refashioned SEU theory into a descriptive theory of decision making, called *Prospect Theory*. Prospect theory proposes a two stage model for decision making under risk. In the initial stage the decision problem is *edited* into a simpler representation. Outcomes, for example, are coded as gains or losses from some reference point. Identical outcomes across alternatives may be cancelled out, and identical outcomes within an alternative may be combined. In the second stage the edited representation is *evaluated* through the application of a value function that captures the empirical observation that people dislike losses more than they like gains of an equivalent amount. The theory accurately predicts that people may choose differently when exactly equivalent problems are worded as gains or losses.

This kind of "framing effect" has been most neatly demonstrated by Kahneman and Tversky in their "Asian disease problem". Subjects are told to imagine that the USA is preparing for the outbreak of an unusual disease, which is expected to kill 600 people. Two alternative programs for combating the disease have been proposed and they should assume that the exact scientific estimate of the consequences of the programs are as follows:

*If program A is adopted, 200 people will be saved (72%)*

*If program B is adopted, there is a one-third probability that 600 people will be saved and a two-thirds probability that no people will be saved (28%)*

The numbers in parentheses represent the proportion of respondents choosing each option and indicate that most people preferred option A, which could be characterised as a riskless option as it guarantees that 200 people will be saved. In comparison, the possibility of loss for option B makes the gamble unattractive.

However, other subjects were presented with the same scenario and asked to choose between option C and option D.

*If program C is adopted, 400 people will die (22%)*

*If program D is adopted, there is a one-third probability that nobody will die and a two-thirds probability that 600 people will die. (78%)*

Of course, programs C and D are actually the same as programs A and B except that now the outcomes are "framed" in terms of the numbers of lives that might be lost. We can see that now the risky option (D) is more popular than the riskless option.

Note that effects of this kind are not confined to naive subjects; experts are susceptible too. McNeil, Pauker, Sox, and Tversky [82] found that both physicians and lay people altered the choice of surgery or radiation therapy as a treatment for lung cancer, depending on whether likely outcomes were described in terms of survival rates or mortality rates.

### 2.1.3. The prescriptive approach

Where normative approaches specify how to make the best decision (or more generally, solve some problem in the best way) and descriptive approaches capture what people actually do, the prescriptive approach to decision making specifies methods for making acceptably good decisions. This does not necessarily entail teaching people how to use a normative model; rather, it may involve finding a method of decision making that people find easy to use and which provides an acceptable degree of accuracy.

However, there is another way of making judgements under uncertainty, which is to replace the human decision maker with an expert system. In the domain of medical diagnosis the earliest attempt to deal systematically with uncertainty was the MYCIN expert system [Shortliffe and Buchanan 76], where conclusions would be qualified with certainty factors. However as the semantics of the resulting numbers was not well-defined (for example, the numbers are not Bayesian probabilities) their meaning and hence their utility is somewhat questionable. Prospector [Duda, Hart and Nilsson 76] has a mechanism based on an approximate form of Bayes' theorem for assigning probabilities to conclusions. However, there are severe difficulties with applying probabilities to uncertain knowledge. The key problems encountered in applying a Bayesian framework are linked to the typical context of decision; for example, in the medical domain, diagnoses can involve several diseases. Traditional Bayesian schemes requiring mutual exclusivity of hypotheses (for example that each patient has only one disease) and conditional independence of evidence are not appropriate as a solution because these assumptions are often unrealistic [Henrion 90].

Furthermore, as we have already seen, the application of numerical probabilities may tend to disguise the fact that those probabilities are derived from only a small amount of relevant evidence. The expert system that we discuss in this paper overcomes this problem by taking a qualitative approach to uncertainty, but also provides quantitative information where it is appropriate to do so. Moreover, the system is primarily a *decision support system*: it does not replace the human decision maker nor specify what the user must do, but it does provide the user with a judgment and, more importantly, it provides the arguments underlying that judgment. This system is known as StAR, an acronym for *St*andardised *A*rgument *R*eport. We regard the use of the StAR system as an important development, particularly because there seems to be an increasing recognition of the importance of *reasons* in reasoning and decision making. An effective decision support system is likely to be one that takes advantage of the way people naturally process information. Therefore, before discussing StAR in detail, we shall briefly look the role played by explicit reasons when making decisions.

## 2.2    The role of reasons in reasoning

Although there has been a tremendous amount of research into the psychology of judgment and choice, psychologists have paid relatively little attention to the role of reasons in influencing judgment and choice behaviour. This may be due to the influence of normative models; in the case of prospect theory the assumption that people are attempting to maximise SEU may have led to a focus on the elucidation of peoples' value functions. Perhaps an enduring distrust of introspection, together with the observation that people sometimes give inaccurate reports on the causes of their behaviour [Nisbett and Wilson 77] has also contributed to a neglect of the study of reasons as an influence on behaviour. In any event, the tide seems to be turning and there is an increasing interest in the study of reasons and also in the notion that human thought is essentially argumentative [e.g. Billig 87].

For example, Curley, Browne, Smith, and Benson [95] recorded "think-aloud" protocols from subjects as they made a series of probability judgments. From these they were able to discern the type of arguments underlying people's judgements - mostly causal ones as it turned out. They also discovered that people only considered a subset of the possible evidence when making their judgments.

Montgomery [83] characterised decision making as a "search for good reasons". and argued that, when faced with a conflict between dimensions, people try to think of a reason to completely ignore one. This is because single minded arguments are cognitively easier. This results in justifications for actions that can be summarised in statements such as "A good soldier must always obey his commander" (even when told to commit crimes?) or "The wishes of the islanders are paramount".

More recently, Shafir, Simonson and Tversky [93] proposed that decision makers often seek and construct reasons in order to resolve conflict, and to justify their choice to themselves and others. This view of decision making predicts several interesting phenomena which Shafir et al review. For instance, Tversky and Shafir [92b] asked some student subjects whether they would like to purchase an attractive vacation package in Hawaii at a special offer price that expires tomorrow, or whether they would take the option of paying a non-refundable fee to hold open the special offer for a further day. One group of subjects were also told that they had either just failed or passed an examination (failure would entail taking a re-sit after the holidays). Regardless of whether they passed or failed, over 50% of subjects said they would purchase the holiday. However, another group of subjects were told that they would be receiving their examination results tomorrow. Of these subjects 61% elected to pay to defer their decision.

From a normative point of view it is hard to make sense of this finding. If most subjects prefer to take the holiday regardless of passing or failing their exams, then most subjects should also take the holiday even when they don't know the outcome of their exams. From a reasons-based analysis, however, these results are more explicable. It is likely that subjects who fail their exams decide that they deserve the holiday as a consolation, whereas subjects who pass the exams decide they deserve it as a reward. But for subjects who do not know their exam outcome, there is no obvious justification for the immediate purchase of the holiday.

In the holiday problem, it seems that subjects in the 'uncertainty' condition fail to fully analyse their situation: they do not attempt to imagine, or are unable to imagine, how they would feel upon the possible outcomes of their exam. It is possible that if subjects had attempted to externalise the arguments for and against taking a holiday they might have reached a rather different decision. There is an abundance of evidence that shows a simple consideration of pro and con arguments to be an excellent way of making good decisions. This approach was described by Benjamin Franklin in a letter to Joseph Priestley in 1772[4]:

"I cannot, for want of sufficient premises, advise you what to determine, but if you will please I will tell you how.... My way is to divide half a sheet of paper by a line into two columns; writing over the one Pro, and over the other Con. Then, during three or four days' consideration, I put down under the different heads short hints of the different motives, that at different times occur to me for or against the measure. When I have thus got them all together in one view, I endeavor to estimate the respective weights...[to] find at length where the balance lies....And, though the weight of reasons cannot be taken with the precision of algebraic quantities, yet, when each is thus considered, separately and comparatively, and the whole matter lies before me, I think I can judge better, and am less liable to made a rash step; and in fact I have found great advantage for this kind of question, in what may be called moral or prudential algebra."

Franklin's method simply involves listing positive and negative reasons for a course of action, then assigning a score of +1 to each positive reason and a score of -1 to each negative reason. The reasons are then given intuitive importance weights, and the course of action with the highest resulting sum is the more desirable of the two.

For many important decisions involving psychologically incomparable reasons for action, methods based on weighted averages have been shown to be more accurate than intuitive global judgment [see Dawes, 88], despite the fact that many people place great faith in such judgments. In areas such as the prediction of college success, parole violation, psychiatric diagnosis, physical diagnosis, and business success/failure, researchers have sought to compare the statistically derived weighted averages of the relevant predictors - known as linear models - with the intuitive global predictions of trained experts. For example, if predicting the chances of survival from major surgery, relevant predictor variables may be the age, weight and general fitness of the patient. The linear model is constructed in such a way as to maximise the statistical relationship between the predictor variables and the criterion to be predicted. The value of each of the predictor variables is differentially weighted according to the strength of its diagnostic relationship to the criterion and then all the variables are summed.

Meehl [54] summarised the results of approximately twenty studies in which clinical judgment was compared with the statistical method, finding that in all studies the latter provided more accurate predictions or the two methods tied. Sawyer [1966] obtained the same results in a further forty-five studies, including

---

[4] From Bigelow, J. [1887: p.522], quoted in Dawes [1988: p.202].

two where the clinical judge had access to more information than did the statistical model, yet still did worse!  Now in these studies, the linear models used statistically optimal weights.  However, Dawes [79] investigated the possibility that *any* linear model might outperform clinical judgments.  He examined five different situations where clinical judgments could be compared with statistical models.  These included psychiatric diagnoses of psychosis as well as college performance at two different Universities.  He constructed 20,000 "improper" linear models by determining weights randomly (keeping the sign of the cue [it's direction of influence] correct), and found that, overall, the "improper" linear models were more successful at making these predictions than were the "experts" - ranging from graduate students to eminent clinical psychologists.

The significance of this finding is that the superiority of statistical models is not critically dependent on the values of the weights attached to each of the cues but simply the cues themselves.  Unlike human judges statistical models will not be influenced by fatigue and boredom or distracted by spurious factors to occasionally discount cues inappropriately.  Human judges have been described as "cognitive misers" who seek excuses to ignore potentially valid information in the interests of reducing cognitive strain.  The inability of experts to outperform statistical models - even those with random weights - can therefore be attributed to their inconsistent focusing on selected cues and their consequent failure to appropriately integrate all the available information.  Statistical models will always give some weight to all the cues that are present in the model.  As Dawes and Corrigan [1974] commented: "the whole trick is to know what variables to look at and then know how to add."

We have seen, then, that people tend to consider reasons when making decisions.  As Dawes [88] points out, decisions can be very inaccurate if based on an intuitive judgment, but where people's reasons are explicitly stated and weighted as to their importance, it is possible to reach a very high standard of decision accuracy.  However, many domains of interest, as discussed above, involve uncertainties arising from lack of information.  Despite this, it is still possible to make judgments from such information as is available, as long as those judgments reflect the uncertainty inherent in the information.  The expert system (StAR) we shall describe can make such judgments based on limited information and make transparent to the user the reasons behind the decision.

 The current version of StAR has been specifically designed for use in the domain of toxicological risk assessment.  We have already seen that the nature of risk assessment in this domain is often problematical.  However, there are also other reasons why decision support systems may be useful in this area.  For example, experts may be subject to particular biases in their reasoning.  Carlo, Lee, Sund and Pettygrove [92] conducted a study where a large sample of epidemiologists, toxicologists physicians and general scientists received information about three substances. Each subject was read a brief vignette written to reflect the mainstream scientific thinking about one substance.  For half of the respondents this substance was referred to only as substance X, Y or Z.  The other respondents were told that the substance was either dioxin, radon or environmental tobacco smoke (ETS).  Revealing the name of dioxin didn't have any significant effect on health risk assessments but the experts were significantly more likely to rate radon and ETS as a serious environmental health hazard when they were told the name of the substance. Carlo et al conclude that experts' evaluations of scientific data may be influenced by values and experiences which might in turn bias estimates of risk.

A second problem for expert judgments about risk concerns the non-homogeneity of certain experts as regards their opinions about risks in their domain. A study of toxicologists and lay persons' judgments of chemical risks by Malmfors, Slovic, and Neil [93] showed substantial disagreement among the experts, as well as between the experts and the public. The toxicologists were particularly divided about the ability of animal tests to predict a chemical's effect on human life. Toxicologists working in industry saw chemicals as more benign and were somewhat more confident than their counterparts in academia and government in the general validity of animal tests - except when those tests provided evidence for carcinogenicity, when many of the industrial experts changed their minds. Malmfors et al argue that these findings suggest that controversies over chemical risks may be fuelled as much by the limitations of risk assessment and disagreement among experts as it is by public misconceptions.

One way of dealing with the lack-of-consensus problem among experts may be to accept it as a fact of life. What is important is that risk communications - whether to other experts, the government or the public - make quite clear the reasons that lie behind the judgement.

# 3.    The StAR risk adviser

## 3.1    The StAR approach

Because of the judgmental nature of much medical reasoning it has been necessary to develop a framework for decision making which permits use of qualitative methods for managing the uncertainty in medical judgments as well as quantitative inference. The basic idea is straightforward; the method involves the construction of "arguments" which are, minimally, "for" or "against" propositions. In general, arguments are represented as triples comprising the claim which is being considered, the grounds of the argument, and some symbolic or quantitative representation of the "force" or strength of the argument. Following a proposal from Fox, [Fox and Clarke, 91; Fox 94] the StAR approach explores the use of argumentation in risk assessment. The basic idea is related to Toulmin's [58] scheme for representing everyday arguments. In order to arrive at an assessment of the overall plausibility of a claim the collection of arguments concerning that claim must be aggregated.

One qualitative aggregation strategy is to see if it is possible to establish specific patterns of argument as qualitative landmarks and, if so, to associate those patterns with linguistic terms in a way which has "cognitive validity"; that is, where the definitions in some way reflect people's intuitive usage of the associated terms. These terms can be defined using Prolog rules such as the following, where "supporting" indicates that the argument is pro the claim, and "eliminating" indicates that the argument rules out the claim:

```
possible(Claim) if
        argument(Claim,Argument1,supporting)
        and not argument(Claim,Argument2,eliminating).
```

In other words a claim is possibly true if there is an argument in its favour and no argument excluding it, where Claim is a sentence in some formal or natural language. Note that this definition is domain-independent; it makes no commitment about the kind of arguments that can be used in a particular domain and therefore has the potential for application in multiple domains. Similarly intelligible, and apparently natural, interpretations can be given to a wide range of English uncertainty terms, such as "suspected", "plausible", "equivocal" etc. [Fox 86; Elvang-Goransson et al 93].

In this particular case the two contradictory arguments would leave us "equivocal" about the potential toxicity of the compound. If, however, there is a basis for assigning a quantitative strength to an argument, such as a conditional probability, then this can be substituted for the symbolic qualifiers "supporting/opposing".

The basis for these developments is a well-founded extension to classical logic which permits reasoning about support and doubt [Fox, Krause and Ambler 92]. This yields a novel "Logic of Argumentation" (LA). LA is embodied in a theorem prover which is sound and complete with respect to the semantics of the logic [Krause, Fox and Judson 95]. Various aggregation schemes, including those described, have been justified and implemented, and a scheme for linguistic aggregation of arguments has been defined. The current version of StAR allows two types of weightings for arguments, those that involve 'supporting' evidence and those involving 'strong' evidence ('received wisdom'). Conclusions in StAR are arrived at by the following combination of arguments (or lack of arguments):

| | |
|---|---|
| **Confirmed** | known human carcinogen |
| **Possible** | hazard identification |
| **Probable** | at least one strong argument for and none against (Strong evidence for) |
| **Plausible** | at least one supporting argument for and none against |
| **Equivocal** | at least one supporting argument for and at least one supporting argument against |
| **Doubted** | at least one supporting argument against and none for |
| **Improbable** | at least one strong argument against and none for (Strong evidence against) |
| **Open** | StAR knows nothing |
| **Contradictory** | Strong supporting evidence and strong opposing evidence |

An argumentation framework provides additional advantages which are relevant to the assessment of risk. First, as we would intuitively expect, arguments can be grounded in a diverse range of evidence types, ranging from observations and opinions to information extracted from external databases, from simulations or from logical reasoning [Fox and Clarke 91]. Argumentation provides a common formalism for interpreting diverse information sources as supporting or opposing a claim, without forcing unrealistic assumptions about the precise probability of the evidence being accurate or the argument being valid. The practical benefits of this are that we can combine evidence and arguments of very different kinds into an assessment (quantitative, qualitative or linguistic) of the persuasiveness of the claim, together with a structured presentation of the arguments for and against it.

## 3.2.    A demonstration of the StAR risk adviser

Having described the rationale behind StAR, some examples of StAR in use will now be given. These are taken from the current StAR demonstrator, which has been constructed for the specific purpose of predicting the potential carcinogenic risk due to novel chemical compounds. This is part of a collaborative project involving the Imperial Cancer Research Fund, Judson Consulting, LHASA UK Ltd., Logic Programming Associates Ltd., and City University. Further details can be found in Krause, Fox and Judson [95]. Furthermore, it is anticipated that the StAR approach will be applied to many other domains besides this one. As StAR is still undergoing development the examples presented here, although as realistic as possible, are primarily for illustration.

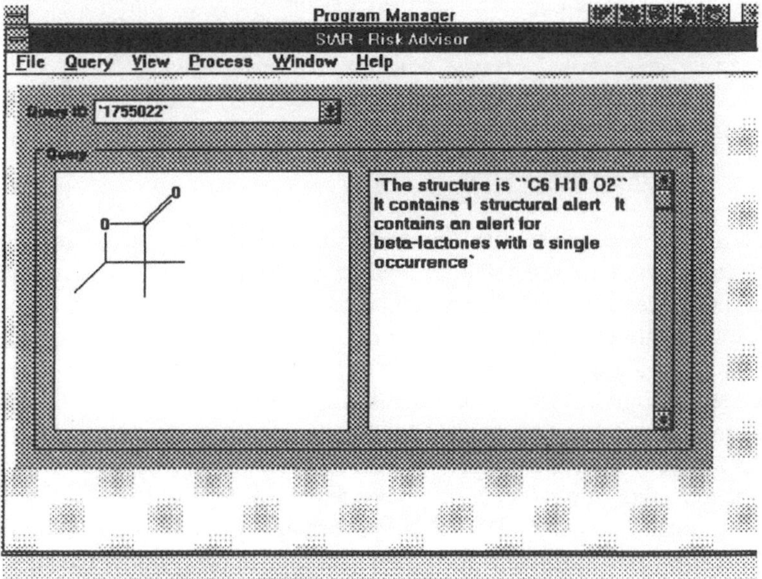


Figure 1. The initial identification of a hazard

Initially, the user of StAR begins by providing the system with the chemical structure of the compound to be assessed, together with any relevant additional information (e.g. exposure routes, the species of animal that will be exposed to the

150

chemical). StAR then searches its database of structural alerts (chemical sub-structures within molecules) to see whether there are any that match part of the entered structure. Figure 1 shows the initial identification of an alert. The left panel of the figure shows the structure that has been entered by the user, and the right panel shows the alert that is reported by StAR. At this point the user can ask for a summary of the alert, shown in Figure 2.

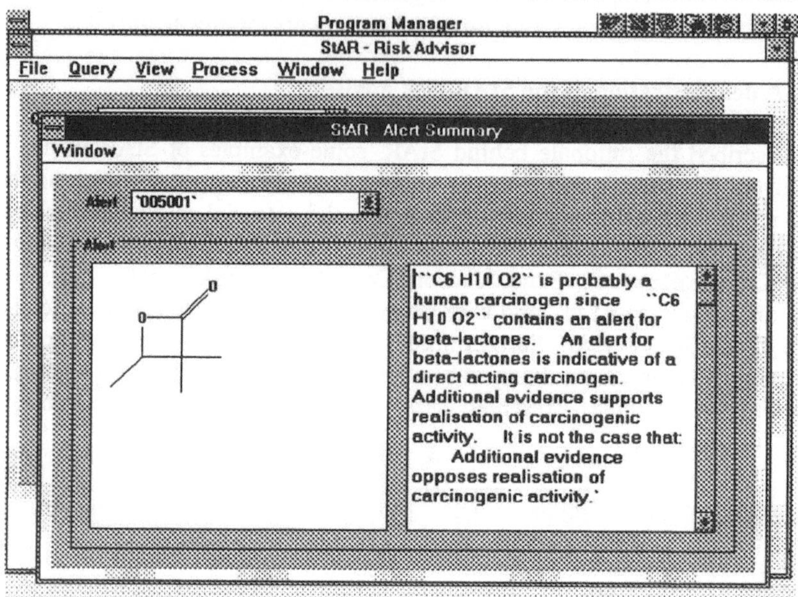


Figure 2. The summary of the alert

When StAR finds a match, as in the example shown here, then a theorem prover attempts to construct arguments for or against the presence of a hazard in the context under consideration. These arguments are presented in a detailed report to the user, together with a summary statement of the risk. The detailed report for this structure is shown in Figure 3.

As can be seen in Figure 3 the report begins with the statement that "C6H10O2 is probably a human carcinogen since..." The report goes on to state that an alert for beta-lactones has been identified, and that beta-lactones are an indicator of a direct acting carcinogen. The empty panel on the bottom left of Figure 3 means that StAR can find no arguments against C6H10O2 being carcinogenic, but in the bottom right hand panel there are two arguments *for* it being carcinogenic. Firstly, the substance may be genotoxic, meaning that the hazard may cause genetic mutation. Secondly, the substance may be carcinogenic to non-humans. Substances that are carcinogenic to non-humans are taken to be indicative of potential carcinogenicity to humans, unless there is evidence to counter this. For example, if a substance only causes cancer in a specific site in non-humans, and humans do not possess such a site, then this is an indicator of non-carcinogenicity in humans. However, this does not apply

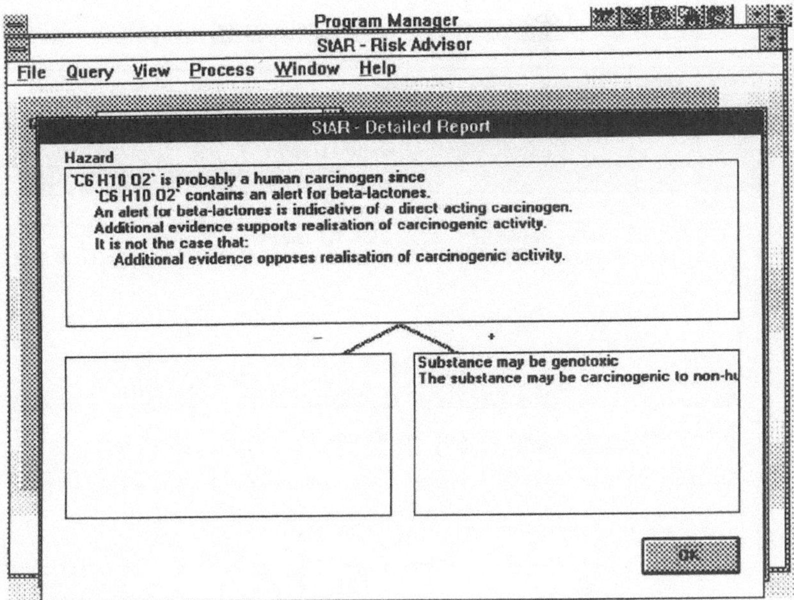

Figure 3.  The detailed StAR report

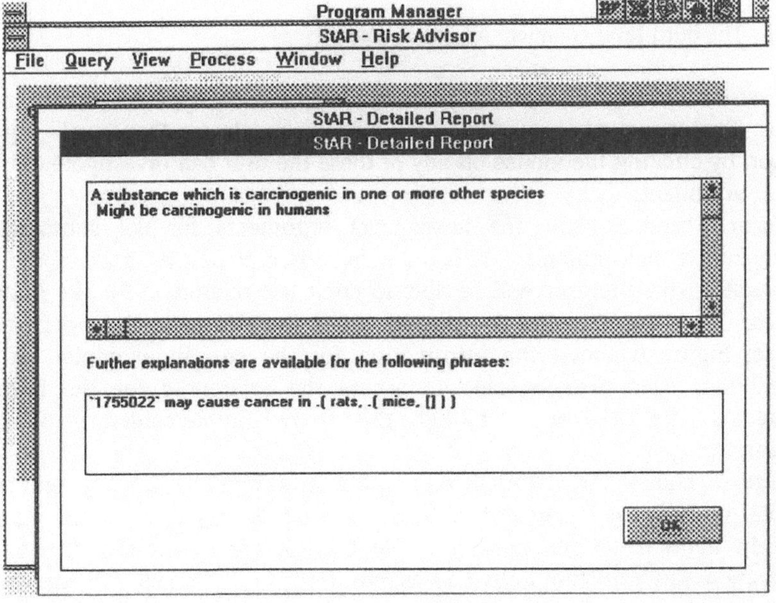

Figure 4.  The second-level argument for carcinogenicity

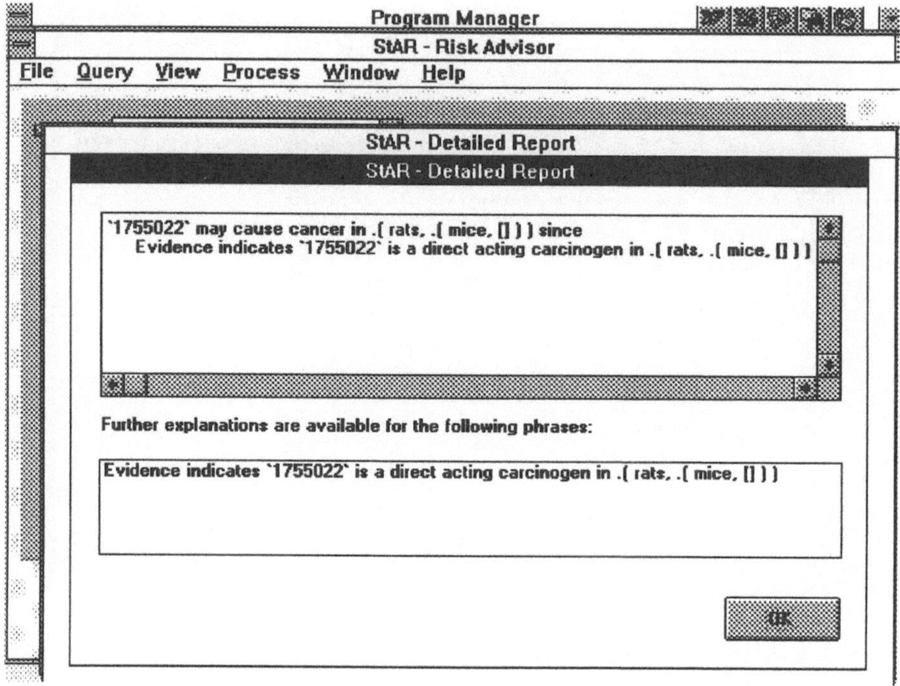



Figure 5.  The third-level argument for carcinogenicity

in the current example where there is no evidence to oppose carcinogenicity in humans. The arguments provided in the bottom panels are "top-level" arguments. However, by clicking the mouse on any of these the user can investigate the backing for these arguments.

Figures 4 and 5 show the lower-level arguments for the substance being carcinogenic to non-humans.  It is intended, though not yet part of the current demonstrator, that the user will be able to elicit the references for the information provided.  It is possible for the user to elicit the information held in the alert database; Figure 6 shows the information for the current example.  It is also intended that users will be able to access the references for the information presented in the risk report, though this is not implemented in the current demonstrator.

Figure 7 shows the detailed report for a different example, the structure C12H6C14.  Here the evidence for human carcinogenicity is stated as "equivocal". The right hand panel indicates that the risk of carcinogenicity to non-human animals increases with cumulative doses, but the left hand panel indicates that there is unlikely to be any effect on genetic structure, and the substance is unlikely to be mutagenic.

It is the explicit representation of the arguments underlying risk assessments that is the central innovation of StAR.  Qualitative risk assessments of novel chemical structures are generated however, quantitative information can be exploited where

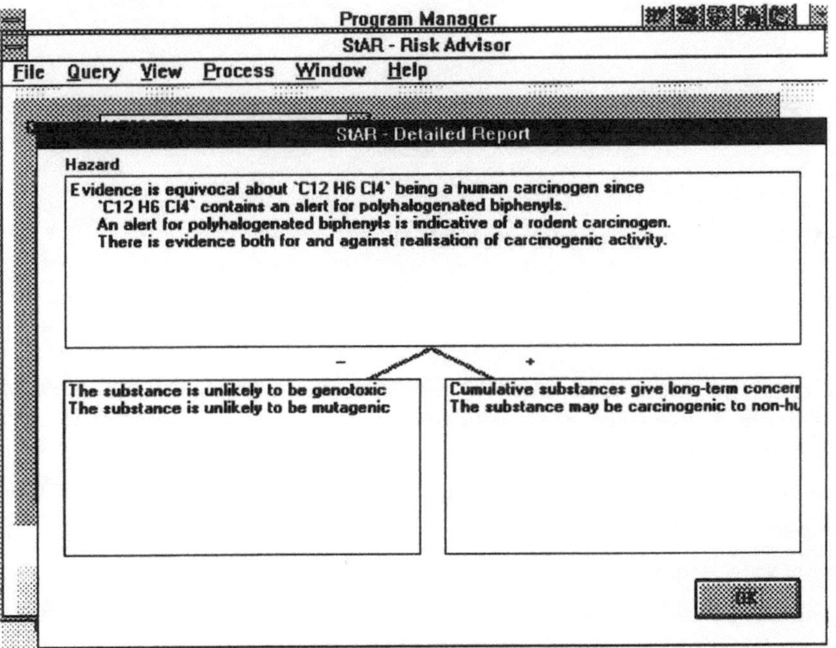


Figure 6.  Information from the alert database



Figure 7.  An example of an "equivocal" report

appropriate. Figure 8 shows a report for a chemical structure where the evidence for the presence of carcinogenicity is offset by the evidence against in the form of the structure's logP value. The logP value gives an indication of how easily the chemical is absorbed into fatty tissue. In this case, the value of 2.4 indicates that the chemical is unlikely to be absorbed.

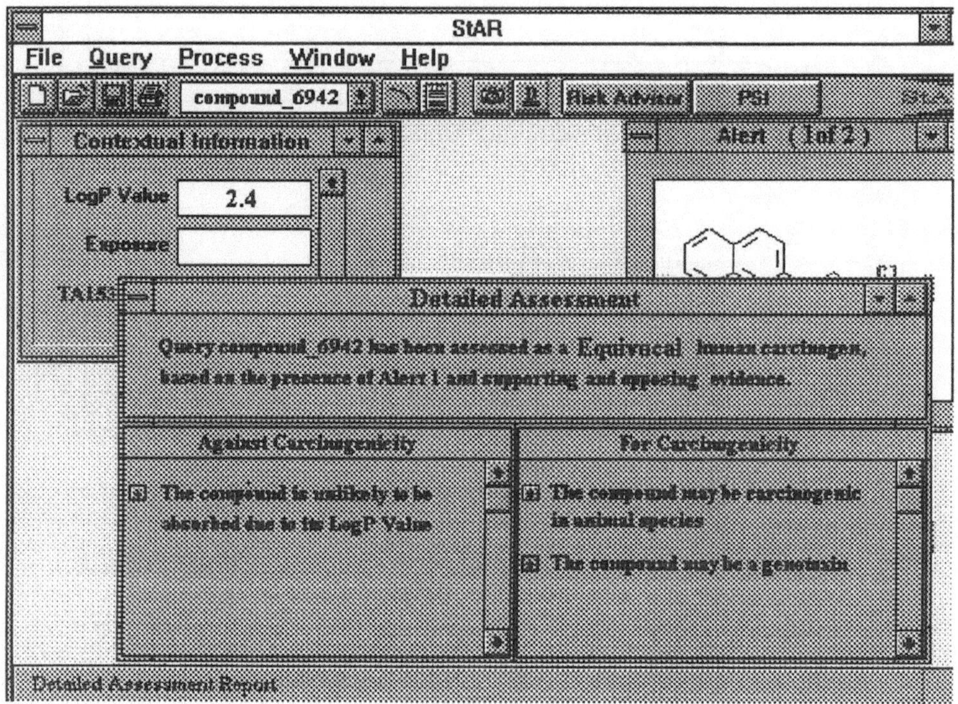


Figure 8.  The use of quantitative information in StAR

# 4. Discussion

- The current application has been designed for toxicological risk assessment, where the problems of assessment can be summarised as follows [Krause et al, 1995 discuss these in more detail]:

- toxicological risk assessments for chemicals may at best cover a very wide range of possible values;

- point value estimates conceal the uncertainties inherent in risk estimates;

- judgments based on the comparison of point values may be quite different from those based on the comparison of ranges of possible values;

- in very many cases the spread of possible values for a given risk assessment may be so great that a numerical risk assessment is completely meaningless.

These problems of risk assessment are not confined to toxicology, however. The StAR approach provides a generic solution for reasoning under uncertainty and it is anticipated that the StAR approach will be applied to other domains. The chief benefits offered by StAR are that it not only provides a qualitative assessment of risk based on a well-founded Logic of Argumentation, but that it provides the user with the arguments underlying the decision. This may be particularly useful for the decision maker who has to account to some third party for the decision taken, or who has to recommend a course of action. At the very least, making transparent the reasons behind StAR's conclusions allows the user to see whether a risk assessment is based on a consideration of much evidence, or whether the underlying evidence is piecemeal.

Currently, work is being undertaken to determine the most appropriate terms to be used as qualifiers in the summary statements. A considerable amount of research has previously been conducted into the meaning of words such as "probable" and "possible" (for a review see Wallsten and Budescu, 1995], yet most (not all) such work has concentrated on elucidating numerical meanings for such words. We are interested in how people use such words to summarise arguments. For example, we are about to investigate the kind of linguistic expressions that toxicologists actually use to summarise the kind of arguments that are presented in the StAR risk adviser. In this study, toxicologists will be presented with arguments for and/or against the presence of carcinogenic risk to humans, but with no information about the particular chemical structure of concern.

We also intend to evaluate how StAR is actually used. For example, if people are presented with the same arguments that StAR presents will they come to the same conclusions as StAR? To what extent will users agree with the conclusions that StAR reaches? Will users search for information beyond the top-level arguments and, if so, how deeply?

In summary, the StAR approach is an attempt to provide decision support in assessing risk in domains where relevant background information may be limited. Moreover, by providing users with the arguments underlying risk assessments, StAR operates in a way which is geared closely to the way in which people make judgments. That is, rather than by making explicit computations of numerical probabilities and utilities, it is the reasons for making a particular judgement that are important. Thus, it is expected that users will find the StAR system particularly easy to use and understand. Moreover, our review of the psychology of reasoning and judgement leads us to the view that StAR's reasons-based approach using the logic of argumentation will empower people to make better judgments under uncertainty than they might otherwise be capable of making.

# 5. Acknowledgements

The work reported here is funded under the DTI Intelligent Systems Integration Programme and by the Engineering and Physical Sciences Research Council. We would very much like to acknowledge the work of the StAR team who have been responsible for the development of StAR: John Fox, Philip Judson, Paul Krause, Jan Langowski, Nicola Pappas, Mukesh Patel, Clive Spenser, and Christian Tonnelier.

# 6. References

[Ayton and Pascoe 95] Ayton, P., and Pascoe, E.  Bias in human judgement under uncertainty? The Knowledge Engineering Review, 1995, 10(1), 21-41.

[Bigelow 87] Bigelow, J. (Ed.) The complete works of Benjamin Franklin.  New York: Putnam, 1887.

[Billig 87] Billig, M.  Arguing and thinking: A rhetorical approach to social psychology. Cambridge: Cambridge University Press, 1987.

[Bolger and Wright 94] Bolger, F., and Wright, G. The quality of expert probability judgement: Issues and analysis. Expert Systems, 1994, 11(3), 149-158.

[Carlo, Lee, Sund  and Pettygrove 92] Carlo, G.L., Lee, N.L., Sund, K.G. and Pettygrove, S.D. The interplay of science, values and experiences among scientists asked to evaluate the hazards of Dioxin, Radon and Environmental Tobacco Smoke. Risk Analysis, 1992, 12, 37-43.

[Carter 91] Carter, R.L. (Chairman). Guidelines for the evaluation of chemicals for carcinogenicity.  Department of Health Report on Health and Social Subjects 1991, 42.

[Curley, Browne, Smith and Benson  95] Curley, S.P., Browne, G.J., Smith, G.F., and Benson, P.G.  Arguments in the practical reasoning underlying constructed probability judgments. Journal of Behavioral Decision Making, 1995, 8, 1-20.

[Dalkey 72] Dalkey, N. An elementary cross impact model. Technological Forecasting and Social Change, 1972, 3, 341-351.

[Dawes 79] Dawes, R.M.  The robust beauty of improper linear models.  American Psychologist, 1979, 34, 571-582.

[Dawes, 88] Dawes, R.M. Rational choice in an uncertain world.  Orlando: Harcourt Brace Jovanovich, 1988.

[Dawes and Corrigan 74] Dawes, R.M. and Corrigan, B. Linear models in decision making, Psychological Bulletin, 1974, 81, 95-106.

[Duda, Hart and Nilsson 76] Duda, R.O., Hart, P.E., and Nilsson, N.J. Subjective Bayesian methods for rule-based inference systems.  Proc. Nat. Computer. Conf (AFIPS), 1976, 45, 1075-1082.

[Edwards  54] Edwards, W. The theory of decision making.  Psychological Bulletin, 51(4), 1954, 380-417.

[Edwards 92] Edwards, W. Utility theories: Measurement and applications.  Boston: Kluwer Academic Publishers, 1992.

[Elvang-Gøransson, Krause and Fox 93] Elvang-Gøransson, M., Krause, P., and Fox, J. Dialectical reasoning with inconsistent information.  In D. Heckerman and A. Mamdani (Eds.), Uncertainty in artificial intelligence. San Mateo, CA: Morgan Kaufmann, 1993.

[Feynman 88] Feynman, R.P. . 'What do you care what other people think?':
Further adventures of a curious character. London: Unwin Hyman Ltd, 1988.

[Fischhoff, Slovic and Lichtenstein 78] Fischhoff, B., Slovic, P. and Lichtenstein, S. Fault trees: Sensitivity of estimated failure probabilities to problem representation. Journal of Experimental Psychology: Human Perception and Performance, 1978, 4, 330-344.

[Fox 86] Fox, J. Three arguments for extending the framework of probability. In L.N. Kanal and J.F. Lemmer (Eds.), Uncertainty in Artificial Intelligence. Amsterdam: North-Holland, 1986.

[Fox 94]. Fox, J. On the necessity of probability. In G. Wright and P. Ayton (Eds.), Subjective Probability, Chichester, UK: John Wiley & Sons, 1994.

[Fox and Clarke 91] Fox, J., and Clarke, D. Towards a formalisation of arguments in decision making. Proceedings of Stanford Spring Symposium on Argumentation and Belief, March 1991.

[Fox, Krause and Ambler 1992] Fox, J., Krause, P.J., and Ambler, S.J. Arguments, contradictions and practical reasoning. Proceedings of ECAI '92, John Wiley and Sons, 1992

[Health and Safety Executive 96] Health and Safety Executive. Use of risk assessment within government departments. HSE Books, 1996.

[van der Heijden 94] van der Heijden, K. Probabilistic planning and scenario planning. In G. Wright and P. Ayton (Eds.), Subjective Probability, Chichester, UK: John Wiley & Sons, 1994.

[Henrion 90] Henrion, M. Towards efficient probabilistic diagnosis in multiply connected belief network. In R.M. Olivier and J.Q. Smith (Eds.), Influence diagrams, belief nets and decision analysis. John Wiley & Sons, 1990.

[Kahneman and Tversky 79] Kahneman, D. and Tversky, A. Prospect theory: An analysis of decisions under risk. Econometrica, 1979, 47, 263-291.

[Kahneman and Tversky 81] Kahneman, D. and Tversky, A. The framing of decisions and the psychology of choice. Science, 1981, 211, 453-458

[Krause, Fox and Judson 95]Krause, P.J., Fox, J., and Judson, P. Is there a role for qualitative risk assessment? Paper presented to the Eleventh Annual Conference on Uncertainty in Artificial Intelligence: UAI-95, 1995.

[Lichtenstein and Slovic 71] Lichtenstein, S. and Slovic, P. Reversals of preference between bids and choices in gambling decisions. Journal of Experimental Psychology 1971; 89: 46-55.

[Lichtenstein and Slovic 73] Lichtenstein, S. and Slovic, P. Response-induced reversals of preference in gambling: An extended replication in Las Vegas. Journal of Experimental Psychology, 1973, 101, 16-20.

[Lichtenstein, Slovic and Zinc 69] Lichtenstein, S., Slovic, P. and Zinc, D. (1969). Effect of instruction in expected value on optimality in gambling decisions. Journal of Experimental Psychology, 79, 236-240.

[Malmfors, Slovic and Neil  93] Malmfors, T., Slovic, P., and Neil, N.K. Intuitive toxicology: Expert and lay judgments of chemical risks. Comments Toxicology, 1993, 4(6), 441-484.

[McClelland and Bolger 94] McClelland, A.G.R., and Bolger, F. The calibration of subjective probabilities: Theories and models 1980-1994. In G. Wright and P. Ayton (Eds.), Subjective Probability. Chichester, UK: John Wiley & Sons, 1994.

[McNeil, Pauker, Sox and Tversky 82] McNeil, B.J., Pauker, S.G., Sox, H.C., and Tversky, A. On the elicitation of preferences for alternative therapies. New England Journal of Medicine, 1982, 306, 1259-1262.

[Meehl 54] Meehl, P.E. Clinical versus statistical predictions: A theoretical analysis and review of the evidence. Minneapolis: University of Minnesota Press, 1954.

[Montgomery 83] Montgomery, H. Decision rules and the search for a dominance structure: Towards a process model of decision making. In P. Humphreys, O. Svenson, and A. Vari (Eds.), Analysing and aiding decision processes. Amsterdam: North-Holland, 1983.

[Nisbett and Wilson 77]Nisbett, R.E., and Wilson, T.D. Telling more than we can know: Verbal reports on mental processes. Psychological Review, 1977, 84, 231-295.

[Payne, Bettman, Coupey and Johnson 92] Payne, J.W., Bettman, J.R., Coupey, E., and Johnson, E.J. A constructive process view of decision making: Multiple strategies in judgement and choice. Acta Psychologica, 1992, 80, 107-141.

[Raiffa 68] Raiffa, H. Decision Analysis. Reading, MA: Addison-Wesley,1968.

[Royal Society  92] Royal Society. Risk: analysis, perception and management. London: Royal Society, 1992.

[Sawyer 66] Sawyer, J. Measurement and prediction, clinical and statistical. Psychological Bulletin, 1966, 66, 178-200.

[Shafir, Simonson, and Tversky 93] Shafir, E., Simonson, I., and Tversky, A. Reason-based choice. Cognition, 1993, 49, 11-36.

[Shortcliffe and Buchanan 76] Shortcliffe, E.H., and Buchanan, B.G. A model of inexact reasoning in medicine. Mathematical Biosciences, 1976, 23, 351-379.

[Slovic, 1972] Slovic, P. Psychological study of human judgement: implications for investment decision making. Journal of Finance, 1972, 27, 779-799.

[Slovic, Griffin and Tversky,  90] Slovic, P., Griffin, D., and Tversky, A. Compatibility effects in judgement and choice. In R.M. Hogarth (Ed.), Insights in decision making: A tribute to Hillel J. Einhorn. Chicago, IL: University of Chicago Press, 1990.

[Toulmin 58] Toulmin, S.E. The uses of argument. Cambridge: Cambridge University Press, 1958.

[Tversky and Kahneman 83] Tversky, A., and Kahneman, D. Extensional versus intuitive reasoning: the conjunction fallacy in probability judgement. Psychological Review, 1983, 90(4), 293-314.

[Tversky and Shafir 92a] Tversky, A., and Shafir, E. Choice under conflict: The dynamics of deferred decision. Psychological Science, 1992a, 3, 358-361.

[Tversky and Shafir 92b] Tversky, A., and Shafir, E. The disjunction effect in choice under uncertainty. Psychological Science, 1992b, 3, 305-309.

[Tversky and Simonson 93] Tversky, A., and Simonson, I. Context-dependent preferences. Management Science, 1993, 39(10), 1179-1189.

[Tversky, Sattath and Slovic 88] Tversky, A. Sattath, S. and Slovic, P. Contingent weighting in judgement and choice. Psychological Review, 1988, 95, 371-384.

Wallsten and Budescu [95] Wallsten, T.S., and Budescu, D.V. A review of human linguistic probability processing: General principles and empirical evidence. The Knowledge Engineering Review, 1995, 10(1), 43-62.

[Zeckhauser and Viscusi 90] Zeckhauser, R.J., and Viscusi, W.K. Risk within reason. Science, 1990, 248, 559-564.

# Is there a Role for Third Party Software Assessment in the Automotive Industry?

Roger Rivett
Rover Group Ltd.

## Abstract

The use of third party assessment for software intensive safety related systems is often suggested by standards for higher integrity level systems. A recent European Esprit project, CASCADE, (Certification and Assessment of Safety-Critical Application Development) has devised a new assessment method addressing the certification and assessment of software intensive safety critical systems in the railway and automotive industries. While there has been much interest from the railway industry in the new method, this has not been the case within the automotive industry. This paper describes the automotive environment with regard to market, legislation, safety and validation and explains how these factors affect the use of third party validation for software based systems.

This report reflects work which is partially funded by the Commission of the European Communities (CEC) under the ESPRIT III programme in the area of Information Processing Systems, project no. 9032: "Certification and Assessment of Safety-Critical Application Development".

## 1 Introduction

CASCADE (Certification and Assessment of Safety-Critical Application Development) is a multi-national project funded by the CEC under the ESPRIT III programme. CASCADE is addressing the certification and assessment of software intensive safety critical systems in the railway and automotive industries. [CASCADE]

The principal deliverable of CASCADE is the Generalised Assessment Method (GAM) for assessing software intensive safety critical systems. The GAM is concerned with both process and product evaluation and addresses the following issues:

- Assurance of the correct integrity level of safety critical software.

- The role of a safety case and its contents.

- The types of demonstration (testing, proof, audit, inspection, etc) to be used for the assessment of safety-related software.

- The procedures, methods and techniques for software assessment.

- Common criteria and rules to adopt.

- Application proof of conformity to the relevant standards for certification purposes.

As the electronic and software content of vehicles is increasing it would seem that there is a need within the automotive industry for such a method. However it is apparent that there is less interest in the work of CASCADE in the automotive industry than in the railway industry.

This paper seeks to present an overview of the automotive industry and explain the legal, marketing, and design environment within which it operates. In particular the following topics are discussed:

(i) The legal framework under which the automotive sector operates.

(ii) Road transport safety issues.

(iii)The general nature of the automotive market.

(iv) The software content of vehicles.

(v) The approach to software development and vehicle validation used in the automotive industry.

From this analysis the current and future roles of third party software assessment within the automotive industry is discussed.

# 2 The Law (UK Only)

The first UK Road Traffic Act was passed around 1900. In subsequent years more acts were passed and by 1978 the whole of the UK automotive industry was governed by the Construction & Use Regulations and the Road Vehicle Lighting Regulations, two very slim volumes. Up to 1978, except for buses and London black cabs, the legislation applied to the driver of the vehicle, not the manufacturer.

## 2.1   1963: UN ECE 1958 Agreement

In 1963 the UK signed the UN ECE 1958 Agreement on motor vehicle equipment and parts.

The Agreement was produced in 1958 by the United Nations Economic Commission for Europe, based at Geneva, in order to facilitate free trade in Europe. It was considered necessary because many countries had produced road traffic laws which, while broadly similar, were not identical and not recognised outside of their own national boundaries. This meant that motor manufacturers had to produce different versions of their vehicles for each different country.

The Agreement, which relates to parts and equipment for motor vehicles, is based on reciprocal recognition by signatory countries of approvals of motor vehicle parts included in the Agreement.

Since 1958 more and more countries have become signatories and the scope is European in its widest geographical sense, e.g. includes Russia and former communist countries. Currently there are about 28 signatories.

The agreement now includes approximately 100 different regulations. The parts and systems covered are wide-ranging and include lights, noise, radio interference, locks, exhaust emissions, seats, brakes and even motor-cycle riders' helmets and advance warning triangles. Some of them concern motorbikes, trucks or buses.

## 2.2 1978: Type Approval (Great Britain) Regulations

As a result of becoming a member of the European Economic Community in 1973, in 1978 the UK introduced a type approval system for passenger cars; Type Approval (Great Britain) Regulations 1978.

This was necessary because in 1970 the Council of the European Economic Community adopted Directive 70/156/EEC - Type Approval of Motor Vehicles and their Trailers. This is a framework directive and references 44 partial directives. These partial directives are concerned with particular aspects of the vehicle and are similar to the individual regulations of the UN ECE Agreement.

Although 70/156/EEC is completely separate from the UN ECE Agreement, the two schemes do try to stay technically in step with each other.

The EEC created its own system in order that eventually it could become a means to facilitate free trade in new motor vehicles and new components for motor vehicles between the Member States of the Community.

From 1970 EEC Member States had to recognise type approvals of vehicle systems or components that were granted to any of the partial directives within the Directive 70/156/EEC framework. For political reasons related to the import of vehicles from non EEC countries, the total list of partial directives envisaged when directive 70/156/EEC was made were not all adopted. This is because it would have automatically triggered a whole vehicle type approval system with the Community.

To sell in an EEC Member State from 1970 manufacturers of vehicles have had the options of:-

1. satisfying the national requirements of that Member State alone;

or

2. satisfying the national requirements of that Member State, with some of those being substituted by EEC partial Directives or by any UN ECE Regulation which that Member State had adopted.

Notwithstanding that, from 1992 exhaust emissions Directives had become mandatory requirements across the Community.

## 2.3  1992: European Whole Vehicle Type Approval

In 1992 the EEC adopted a Directive amending Directive 70/156/EEC which requires that any mass-produced M1 vehicles (passenger vehicles with up to 9 seats including the driver) type approved for sale within the Community after 1st January 1996 comply with that sub-set of the partial directives that apply to M1 vehicles. From January 1998 this requirement for compliance with European Whole Vehicle Type Approval (ECWVTA) will apply to all new mass-produced M1 vehicles sold and registered in the Community.

## 2.4  Type Approval (Homologation)

Type Approval, also known as homologation, is the process of ratification by a third party that an article, or series of identical articles, meets the legal requirements to which it is subject prior to being placed on the market. The manufacturer, having obtained type approval, is committed to ensure that any further examples of the approved article are built in conformity with the approved type and generally has to attest to each article being in conformity by either applying a special marking to the articles and/or issuing a document, a certificate of conformity, with each individual article.

Another means of having a manufacturer declare that he is manufacturing articles to a particular set of requirements is "self-certification". Here, the manufacturer again attests that the individual articles comply but without there being any prior ratification by a third party. "Self-certification" is the method used in the USA to show compliance with the Federal Motor Vehicle Safety Standards.

It is common, both in markets where type approval applies and where self-certification applies, for the authorities to apply some form of after market sampling to ensure that the manufacturer's attestations of compliance are indeed true.

Type Approval is not unique to the automotive industry, for example it is used within civil aviation and telecommunications. An example of its use in telecommunications is that all telephones have a green stamp stating that the equipment is "approved for connection to telecommunication systems".

Type Approval works by having regulations which apply to a particular component or system and which state the conditions under which that item is acceptable and how acceptance is to be determined. A manufacturer wishing to Type Approve an item under a particular scheme submits a production representative sample to an approved agent who determines whether or not it meets the regulation. Once a manufacturer has had an item Type Approved it can then be manufactured and sold in any country which is party to the scheme.

The advantage of type approval is that it allows manufacturers to design products and easily obtain approval to sell them in other countries, that is it facilitates free trade.

The details of type approval regulations can vary widely, e.g.:

- number of vehicles required for approval process

- whether approval process consists of checks and/or tests

- whether tests are conducted by the manufacturer or the assessor

- whether tests are conducted at the manufacturer's premises or the assessor's

- whether the approval process includes any examination of the manufacturing process

Even if the regulations do not explicitly include any check of the manufacturing process it is implicit that those parts made for production will be the same as those supplied for approval.

Legislation applies at three different levels of detail:

- Some legislation applies to particular components, for example horns, headlights, tyres, seat belts. Typically these components are fitted to many different vehicle models from different vehicle manufacturers. The legislative requirements can usually be met without them being fitted to a vehicle.

- Some legislation applies to systems, for example braking. These systems often contain components which are subject to individual legislation but which now have to meet installation criteria as well.

- Some legislation applies to the whole vehicle, for example exhaust emissions and electromagnetic compatibility (EMC).

The legislation gives details of what is to be assessed, how the assessment is to be carried out and what constitutes a pass.

While some legislation gives quite detailed mechanical specifications, for example tyre minimum tread depth, none of the current legislation specifically mentions electronic components let alone the software they contain.

The system requirements are given and checked at a system level. Details of how the requirements have been met is outside the scope of the legislation.

The future for type approval schemes is for global harmonisation. This will be in the next century and has to overcome quite large global differences, e.g. self certification in the USA.

## 2.5 Conclusion

Although there is much legislation with regard to the automotive industry, the current legislation does not require any third party assessment of software. This may reflect the fact that the legislation has largely been put in place to facilitate trade rather than in response to road transport disasters. This is perhaps in constrast to the legislation in other transport sectors.

## 3 Road Transport Safety

In order to travel safely on the road it is not sufficient to consider only aspects of the vehicle design and construction, but also the design and construction of the road infrastructure, the ability and behaviour of the driver and societal attitudes in general.

Vehicle and road infrastructure design and construction are engineering issues and within these can be seen two distinct categories of possible safety measures. Firstly there are measures taken to avoid the occurrence of accidents, often referred to as primary safety measures; and secondly measures taken to preserve human life and health in the event of an accident, often referred to secondary safety measures.

Driver ability and behaviour together with societal attitudes are human factors issues, and while engineering measures can help ameliorate hazards arising from such issues, they can not address the root causes.

## 3.1 Social Aspects

This relates to general road safety, for example as typified by the British 'Highway Code'. It includes things like the attitude to drinking and driving, the use of seat belts and speed limits. In general, society's attitudes of the day are reflected in legislation enacted by the government. However there is often widespread acceptance of the breaking of some laws, for example speed limits, probably stemming from some un-articulated notion of personal freedom.

## 3.2 Driver Ability/Behaviour

With current vehicles, the driver is the principal controlling agent and the bulk of the responsibility for safety on the roads lies with them. Ensuring that drivers have sufficient competence and experience is achieved by the use of driving tests which are usually enshrined in government legislation. Driver behaviour is affected by legislation, for example requiring "due care and attention", and by societal pressure to conform to what is acceptable. However the driver remains the safety component with the greatest variability in the road transport system.

## 3.3 Road Infrastructure

This aspect includes things like road layout design, road surfaces and road traffic signs. Once again the government is responsible for these via legislation or agencies which inspect completed roads. This appears to be the only aspect of the road transport system which regularly uses third party assessment of some sort.

## 3.4 Vehicle Design and Construction

The vehicle manufacturer takes a whole vehicle approach to safety. This means considering both primary and secondary safety measures, for example:

- primary safety
    - ensuring robustness of design and manufacture
    - provision of good driver visibility
    - provision of good road handling characteristics
    - provision of good steering characteristics
    - provision of good braking characteristics
    - provision of ergonomically designed driver information systems
    - provision of comfortable internal environment

- secondary safety
    - deformation of vehicle body under crash conditions so as to protect the occupants
    - provision of seat belts
    - provision of air bags

In some of these areas the government sets minimum requirements and they are enforced in the ways described above (section 2 *The Law (UK Only)*). However in practice the performance of vehicles far exceeds that required by the regulations due to other pressures for increased safety, described below.

With regard to safety objectives and policies a hierarchical approach is taken. The senior management will set vehicle level objectives. Major sub-systems will then be the responsibility of a chief engineer who prescribes design guidelines and performance criteria for their sub-systems. At a component level the responsibility resides with a component engineer, who will decide the procedures and standards necessary in order to achieve the goals set by their own management. For current systems, the embedded software is contained within a single component and so decisions concerning issues like third party assessment are the responsibility of the component engineer, with help and guidance from company policies, standards and experts.

## 3.5 Pressures For Increased Safety

Although, strictly speaking, vehicle design and construction is only governed by the Type Approval regulations the real standards are set by pressure from other sources.

### 3.5.1 Published reports

These originate from a variety of sources and their purpose is to inform a potential buyer of how different vehicles compare with each other with regard to safety. Many also campaign for particular measures in advance of legislation.

*Which? Reports.* This is a consumer's magazine in the UK. It performs crash tests, feature specification analysis and design analysis in order to give a model a safety rating.

*Automotive Magazines.* These perform their crash tests on similar vehicles and publish comparative results.

### 3.5.2 Insurance companies

These analyse the actual injury related costs on a per model basis and publish the results. The results may affect insurance premiums. In the USA the Insurance Institute of Highway Safety (IIHS) perform their own high and low speed crash tests. Note - current legislation in the USA only requires high speed crash tests.

### 3.5.3 Governments

Some governments produce league tables of vehicle safety based on crash statistics from the police and other sources.

### 3.5.4 New Car Assessment Programme (NCAP)

This originated in the USA in the early 1980s. Since then Australia has introduced it, followed by the UK in 1995. It is likely that there will be a European community wide NCAP by the year 2000.

In the US scheme a government agency performs crash tests on a group of similar products and publishes the results in buyer's guides. The tests performed exceed the legal requirements and include front direct, front offset and side impacts. The performance of both the vehicle body and the restraint systems are assessed. The purpose of NCAP is to push vehicle crash performance standards ahead of legislation.

### 3.5.5 Competitors

Each vehicle manufacturer is keenly aware of what safety features its competitors are offering and so comes under pressure to keep pace in order to maintain the same market image in the mind of its customers.

### 3.5.6 Product Liability

The product liability legislation, which includes the concept of "strict liability", states that meeting a legal requirement does not absolve the manufacturer of their responsibility. Ever conscious of this fact, the vehicle manufacturers strive to ensure that their vehicles are not vulnerable under this legislation.

## 3.6   Conclusion

There are four aspects to road safety: design and construction of the vehicle; design and construction of the road infrastructure; the ability and behaviour of the driver; and societal attitudes in general. Of these, the automotive manufacturer can only address the first.

Manufacturers take safety very seriously and are keenly aware of the external pressures for measures over and above those required by legislation. However these external pressures have more bearing on secondary safety measures than primary safety measures. For software, the primary safety measures relate to the robustness of the design. Whereas responsibility for this generally resides with the component engineer, for future systems, with more and more integration by use of vehicle networks, the responsibility may move upwards from the component engineer to system-level chief engineer.

The obvious weak links in the chain at the moment appear to be the driver, whose poor driving is the cause of the majority of accidents, and the attitude of society at large which tolerates the current level of vehicle accidents due to poor driving.

## 4   The Automotive Market

## 4.1   Market Size

The automotive market is very large, for example the total number of cars sold throughout the world in 1995 was 34,614,000 [Worldcar 96]. The impact of the automotive industry on a country can also be very large, for example it may contribute up to a sixth of a country's gross domestic product.

Most automotive manufacturers operate globally, in terms of both manufacturing and marketing. This is necessary to avoid being affected too much by local economic variations. In general, profit margins are low, say less than 2% on average. High investment is required in manufacturing plant. The lead times for new product introduction are high. In order to help cut costs there are collaborative ventures, for example the use of common components. All these factors combine to make the industry very cost conscious with respect to both product development and sales price.

Although the automotive industry is global, it is currently dominated by the three mature markets of West Europe, North America and Japan who together account for about 75% of all world sales( see Table 1).

| Market | Sales(1000s) | PercentageOf World Market |
|---|---|---|
| West Europe | 11,996.90 | 35% |
| North America | 9,397.10 | 27% |
| Asia (inc. Japan) | 7,149.10 | 21% |
| Latin America | 1,963.40 | 6% |
| East Europe | 1,581.20 | 5% |
| Other | 2,526.40 | 7% |

Table 1: World Cars Sales 1995 [Worldcar 96]

A mature market is one where car ownership is unlikely to increase much, the majority of cars being bought on a replacement basis only. Little substantial growth is expected over the next 10 years, say only 2-3 % per annum.

## 4.2 Market Segments

All manufacturers divide the market into different segments, for example small cars, medium cars, executive cars. The definitions for each segment vary from manufacturer to manufacturer and from market to market but the figures shown in Table 2 for 1995 are indicative.

| Market Segment | Percentage of World Sales |
|---|---|
| Basic | 4% |
| Small | 13% |
| Lower Medium | 23% |
| Upper Medium | 24% |
| Executive | 19% |
| Luxury | 2% |
| Multi Purpose Vehicle | 4% |
| Leisure | 10% |

Table 2: World Production by Market Segment 1995
[Worldcar 96]

## 4.3 Conclusion

Electronic and software systems are usually introduced into the upper medium, luxury or executive segments first, and then migrate into the other segments as they get market acceptance. The higher selling price of the high end vehicle helps cover the costs associated with introducing the new technology. This means that in terms of production volumes the growth of new systems is relatively slow.

Although most aspects of the vehicle have software based systems associated with them (see below) the percentage of vehicles fitted with many of them is small, being restricted to the high end segment vehicles. So future growth for current systems can be expected as they get fitted to lower end segment vehicles and as the overall market grows outside of the mature markets.

# 5 The Product

## 5.1 Current Software Applications

Applications to date have tended to replace or supplement traditional functions, for example see Table 3 (this is not an exhaustive list).

| Powertrain Systems | engine management<br>cruise control<br>transmission control |
|---|---|
| Body Systems | exterior lighting<br>wiper systems<br>central locking<br>security systems<br>electric seat controls<br>electric windows |
| Chassis Systems | anti-lock brakes<br>active suspension |
| Other | occupant restraint systems e.g. air bags<br>instrument pack<br>heating and ventilation<br>radio |

Table 3: Current software applications

Although the use of electronics is obviously increasing rapidly there does not appear to be any documentation recording this growth or predicting future growth. Some simple analysis is attempted in section *4.3 Conclusion* above. One figure much used, but un-attributed, is that by the year 2000 the electronic content of a medium-segment vehicle will account for 30% of its value.

Going hand in hand with this growth in electronics is a growth in the use of software, but again this growth has not been documented.

From Table 3 it will be seen that systems which replace or supplement existing functions within the vehicle are approaching saturation point, i.e. there is little scope for new application areas.

## 5.2  Current Integrity Levels

Most of the current systems do not add new hazards to those which can occur anyhow through mechanical failure. For example vehicles have always exhibited the following failure modes: loss of engine power due to lack of fuel; loss of vision due to loss of windscreen wipers or lights; difficulty in steering due to a burst tyre. Most electronic systems controlling these functions do not introduce new hazards.

None of these could be said to be in the highest safety criticality level. Of the current applications it is perhaps only full authority throttle control which introduces new hazards.

As long as systems are confined to the operation of a single vehicle and primary control remains with the driver, the potential for injury to people is confined to that possible for a small number of vehicles carrying up to four or five people. This puts an upper bound on the integrity level of these systems.

So while current vehicle systems are obviously safety related, in the main they are not at the highest integrity levels and the need for third party assessment is correspondingly low, for example a recent draft of IEC 1508 highly recommends assessment by an independent organisation for safety integrity levels 3 and 4, al Automotive Group McGraw-Hill

[IEC 1508].

## 5.3  Future Trends

While systems have been designed which replace or supplement existing functions, only some of these (e.g. engine management, air bags) are standard fit. The others are only fitted on vehicles sold in the more expensive market segments. So one future trend will be the use of more of these systems on vehicles in more market segments.

A trend which is already well established is the use of networks to connect control systems together. This started with the use of diagnostic buses, e.g. ISO9141, which allow a single point connection between all vehicle systems and the service diagnostic equipment. Other buses, e.g. CAN (Controller Area Network), are being used to pass data between control systems. While the use of buses does not necessarily create new functionality it does add to the complexity of the design.

Now that existing vehicle functions have been incorporated into software, future growth will be in the development of new functions only made possible by the use of electronics and software. Examples include use of local radar linked to the cruise control to ensure the maintenance of safe stopping distances. Many of these are likely to take advantage of the networking of systems together. Some new functions may arise just out of the ability of control systems to communicate.

The area with the largest potential growth is that of telematics. These are distributed systems involving central computers analysing data about traffic flow, weather conditions, and communicating with many vehicles via road side transponders. It may also involve vehicle to vehicle communication with corresponding modification of vehicle behaviour, with or without driver intervention. These type of systems require a major change to the road infrastructure and will be impossible without government initiatives and support. The current use of vehicle radios which automatically change station for road traffic announcements is perhaps a tentative step in this direction. An internet address for every vehicle is perhaps the future nightmare that awaits us!

## 5.4  Future Integrity Levels

Systems which take actions which affect the control of the vehicle without driver intervention are much more likely to be placed in higher safety integrity levels. Similarly, systems which communicate with many vehicles and affect vehicle control, either directly or indirectly, are also likely to be in higher integrity levels.

## 5.5  Conclusion

Although the current vehicle systems are in the main in lower integrity levels, future systems are quite likely to be in the higher levels. Therefore the requirement for third party assessment, while currently small, is likely to grow.

# 6 Software Development

## 6.1 The Developers

Although some vehicle manufacturers develop their own software, most software is written by component suppliers.

There are several tiers of component suppliers with first tier suppliers sub-contracting to second tier suppliers and so on. Component suppliers with a mechanical background who find they now have to incorporate electronic controllers into their products will often subcontract this work to an electronics company who may then subcontract the software development. The number of suppliers involved in software development is still a small percentage of the total component supplier base.

Like the vehicle manufacturers, many of the first tier suppliers are large international companies which manufacture and sell in all major markets and to all major manufacturers. Some have a very wide product portfolio whereas some tend to specialise. Often the company, or a particular division of a company, will work exclusively for the automotive industry. This reflects the specialised nature of the work, the need for close co-operation between manufacturer and supplier, and the size of the market.

The majority of the component suppliers have a mechanical engineering background reflecting the historical nature of the industry. More recently companies have developed electrical and then electronic engineering skills. Software engineering skills are the most recent and the least widespread. Most staff involved in software development are more likely to be electronic engineers than computer scientists. This does mean that they are able to relate well to high speed real-time applications.

## 6.2 Standards

### 6.2.1 ISO9000/TickIT

In common with many other industrial sectors, there is widespread adoption of the ISO9000 quality standard within the automotive industry. Some manufacturers use the straight ISO9000 standard while others have derived sector specific versions of it. Once the vehicle manufacturers started to adopt it they then encouraged their suppliers to adopt it. More recently, in the UK, software suppliers are starting to obtain ISO9000 certification under the TickIT scheme. Overall the introduction of the ISO9000 quality standard has had a beneficial effect on all aspects of the automotive business.

### 6.2.2 MISRA Guidelines

Published in 1994, the MISRA Guidelines (Motor Industry Reliability Association) represent the current definition of what is "best practice" in the UK for automotive software development and have authority due to the large number of companies and organisations which endorse them. They were produced voluntarily due to an industry perception that it must improve the general standard of its software development and anticipate the potential need for self regulation in the future. The development of the Guidelines was initiated in response to the UK Safety Critical Systems Research Programme, supported by the Department of Trade and Industry and the Engineering and Physical Sciences Research Council.

The MISRA consortium is still active in promoting the Guidelines with both workshops and a user's forum. The Guidelines have had a major impact on promoting good software engineering practice in the automotive industry both in the UK and the wider world.

## 6.3 Conclusion

As there is no legislative requirement for the use of particular development procedures current practice is very varied.

With the advent of ISO9000/TickIT and the MISRA Guidelines a more standard approach is increasingly being taken. While some suppliers are undoubtedly very good, these are usually first tier suppliers; some of the second and third tier suppliers have greater scope for improvement.

The growing awareness of software standards will also bring greater awareness of the role of third party assessment.

# 7 Vehicle Validation

The automotive industry is fortunate in being able to perform extensive testing of the finished product which samples the full range of the vehicle's intended application domain. This testing, traditionally known as vehicle validation, is taken very seriously by the automotive manufacturers and a large amount of effort is expended in performing it. Historically software based systems have benefited from this vehicle validation and the good track record of the use of software in the automotive industry probably owes a great deal to this.

This section will give an overview of durability and reliability validation testing, but is not intended to be a comprehensive description. Other tests, which are not going to be elaborated here, are also performed. These are:

- Exhaust Emissions

- Electromagnetic Compatibility (EMC)

- Crash worthiness

It will be seen in what follows that the automotive use of the term "validation" is different to that of the software engineering.

## 7.1 Mechanical Approach to Design & Validation

The automotive industry is by tradition a mechanical based enterprise. Design proceeds by producing mechanical drawings. These were formerly paper based but are now produced by the use of CAD and supplemented by the use of finite element analysis. Designs are subject to review and FMEAs are performed to check for design correctness and to highlight possible reliability problems. Prototype components are made by machining metal directly. Later on press tools are constructed which produce the production parts.

In order to validate the design the following validation exercises are performed:

- Component validation for functionality, durability & reliability

- Whole vehicle validation for durability & reliability

Typical figures for whole vehicle validation for the introduction of a new vehicle are 5 million miles. In any one year, across the whole of its product range, a company may perform 12-14 million miles (new vehicles, facelifts, etc.). These figures show that this technique is taken very seriously.

## 7.2   Test Configuration Selection

It is in the nature of the automotive industry to have many variants of each product. These are generated by:

- Model variants, e.g. 3 door, 5 door

- Trim levels, e.g. base model to top of the range

- Customer options, e.g. air conditioning, manual/automatic transmission

This leads to there being hundreds, if not thousands, of possible combinations for the final product. Obviously it is not possible to validate every possible combination so a choice has to be made. Typically only 5 to 10 different vehicle derivatives will be tested. The choice will include those with the largest anticipated sales volume and those which are worst case in terms of the number of components or complexity.

## 7.3   Test Coverage

For a new model the full validation programme will be performed.

Most models have several model year facelifts during their life. These may be major changes, e.g. different engine fitted, or quite minor relating to the visual appearance only.

The degree to which validation is repeated depends on:

- whether or not it is a carryover component, i.e. already used in production on another model

- the degree to which new technology is being introduced

The powertrain (engine/gearbox/transmission) is considered to be critical and a major change to this is likely to lead to all the reliability tests being repeated.

Other changes are judged on a case by case basis.

## 7.4 Durability Testing

This tests the structural integrity under normal operation and in crude terms can be said to check that the powertrain does not break and that the body does not fall apart.

Vehicles are built to production specification, but not with production parts, i.e. parts not necessarily manufactured using a production process.

This exercise validates the design of the vehicle. This activity is carried out according to company proprietary standards but it typically involves driving a number of vehicles a total of 100,000 miles.

## 7.5 Reliability Testing

In contrast to the durability tests which are looking for gross mechanical failures, reliability testing is aimed at achieving reduced customer dissatisfaction and warranty costs.

This is performed on vehicles built with off-tool parts, i.e. parts manufactured using a production process, and effectively validates the manufacturing process.

There are many different types of test which use company proprietary standards and endeavour to simulate 2 years of use. There are three different categories of test: extremes, typical and specific.

## 7.6 Reliability Testing: Extremes

### 7.6.1 Pavé

This involves driving the vehicle over various rough road surfaces which has the effect of inducing a wide range of chassis and engine vibrations.

### 7.6.2 High Vehicle Speed

This has the effect of inducing the maximum oil temperature and engine stress conditions. This particular test represents the majority of the miles driven for reliability testing.

### 7.6.3 Environment

This involves subjecting the vehicle to extremes of ambient temperature, e.g. -40 degrees C and 50 degrees C, and humidity, e.g. water splash tests.

## 7.7  Reliability Testing: Typical

Great effort is put into matching the test conditions with real customer usage profiles and problems that they experience. This information is gathered by:

- Customer Quality Tracking Surveys
  This involves contacting the customer directly during the first 18 months of ownership. Sometimes it is performed on 5 vehicles which are 5 years old.

- buyer surveys

- market research

- instrumenting customer's vehicles and letting them drive for a week while data logging key vehicle parameters. This is often done in the context of fault finding

- analysing warranty return data

- analysing different market requirements.
  This is done on a model by model basis. Even so there is no set usage cycle for a model but each one has a range. For example a 4x4 vehicle may be used as a mobile office by a builder out in all weathers, road conditions and times, or by a teacher who only drives 5 miles a day back and forth to school.

## 7.8  Reliability Testing: Specific tests

Test specifications are supplied by a component area to test a specific aspect of their component. This may be a new technology or design feature or possibly something which arose out of an FMEA.

## 7.9  Problem Reporting & Resolution

Drivers and technicians who perform tests are required to complete logs at the end of the shift. From the shift logs Problem Reports are raised as necessary and sent to the corresponding component area who will investigate and correct the problem. A Problem Report can not be closed down without a sign-off by the Validation area.

A reliability growth curve is produced for each vehicle development phase. There are company standards giving the expected profile and acceptable production values.

## 7.10 Vehicle Validation in the Future

Greater complexity of vehicle systems will make it harder to test all possible combinations, driving conditions and fault conditions. Therefore all available feedback mechanisms will be used to target the testing most effectively.

The trend will be to use more rig tests, as the component areas do at present, and to focus tests more on customer usage as opposed to tests specified by engineering.

The validation department will be more involved with component areas, participating in design reviews and FMEA exercises, and with manufacturing, performing methods build assessment and failure analysis. This approach will enable validation to be targeted towards the safety-critical aspects of the system.

## 7.11 Conclusion

Validation activities are taken very seriously and will be targeted more closely to critical parts of the systems, including safety requirements. The question is, will this be sufficient on its own as systems get ever more complex?

# 8 Discussion

Although there is much legislation with regard to the automotive industry, the current legislation does not require any third party assessment of software. Anticipated work on Type Approval would appear to be concerned with increasing its breadth rather than its depth, so a major impact on third party assessment is unlikely.

As long as society is prepared to tolerate the current level of accidents due to poor standards of driving, any problems that may occur with existing systems will be lost in the noise. Consumer pressure on manufacturers is unlikely to materialise unless an accident does occur that which is directly attributed to software.

The current vehicle systems are in the main at the lower integrity levels. But if telematic systems come into common use then it is likely that the average safety integrity level will increase. This is because the overall system would then consist of more than a single vehicle, maybe hundreds of vehicles, and the potential for accident severity would start to approach that of civil aircraft or railways.

With the advent of these type of telematic systems even comprehensive validation activities will need additional measures to ensure adequate safety levels.

## 9 Conclusion

Is there a Role for Third Party Software Assessment in the Automotive Industry?

There is a small role at the moment for the highest integrity level systems, for example full authority throttle control.

Where new systems take away more of the primary control from the driver, or connect more than one vehicle together in a manner which affect vehicle control, then the role for third party assessment will increase, possibly becoming mandatory for the worst case automatic or multi-vehicle systems.

The need for third party assessment will be driven by the higher integrity levels involved and the inability of traditional vehicle validation to provide sufficient confidence.

Legislation may follow but this is not obvious at the moment.

## References

[CASCADE] CASCADE Summary.
Ricardo Hetherington (Lloyd's Register).
February 7 1996, ref. CAS/LR/RHZ/R480/2 - Public.

[Worldcar 96] World Car Industry Forecast Report, February 1996.
Data Resource Inclusive, Global Automotive Group
McGraw-Hill

[IEC 1508] Draft IEC 1508 - Functional safety: safety-related systems - June 1995. Ed.1
Geneva: International Electrotechnical Commission
(IEC reference 65A Secretariat 123)

# Initial Safety Considerations for an Advanced Transport System


**M.V. Lowson and C.E. Medus**

**Advanced Transport Group,  University of Bristol**

**Bristol BS8 1TR  England**



## Abstract

This paper provides an introduction to some of the safety issues that need to be considered when designing a Personal Rapid Transit (PRT) system.  Realistic but ambitious target accident rates for PRT are set through a comparative examination of other transport modes.  The rate set meets or exceeds accident rates achieved on other forms of surface transport, and is better than those of the car by a factor of more than ten.  Evidence from existing systems suggests that this is not unreasonable.  Issues associated with the design and safe operation of PRT are discussed in general terms, as a preliminary to undertaking a full safety case.  Introduction of such a system offers the potential for major savings over current casualties due to road transport both for passengers and other road users.


## Introduction

The problems, particularly the environmental problems, of current urban transport are now widely recognised.  The benefits to individuals which have led to the explosive growth of car use have resulted in major difficulties to the population as a whole.  In many cities, these problems have led to the growth of substantial peripheral shopping and office areas, and a potential loss of vitality in the city centre.  Solutions to date such as increased emphasis on public transport, and restrictions on car use in the city centre have not been able to reverse the trends.  Current accident rates due to car transport are increasingly unacceptable.  A new solution to these problems is essential if we are to approach the next century with a transport system which provides real benefits to both the user and the non-user.

Ideas recently developed by the Advanced Transport Group (ATG) at the University of Bristol, have demonstrated that a new form of automatic personal transport can be developed which can compete effectively with the car.  ULTRA, the design proposed by the ATG has arisen from a full multi-disciplinary systems analysis of the best solution for future urban transport.  Overall requirements for the best imaginable form of future transport were specified in the broadest possible terms, identifying the needs of both the user and the non-user. In order to ensure that the new system was markedly better than both the car, and current forms of public transport, major objectives were set for improvement: in emissions and energy a factor of 10 compared to current car/road, elimination of congestion, significant improvement in journey times, and 24 hour availability. Equivalent significant reductions in fatalities and serious injuries are a further important objective, and could provide a major part of the rationale for the new system.

There are several key new factors within the present system engineering approach. The first is consideration of a simultaneous change of the infrastructure and the vehicle. Most current approaches to transport problems attempt to use existing infrastructures, such as current roads or railways. These force continued acceptance of the compromises built into existing systems, and are thus unlikely ultimately to provide an effective solution to transport problems of the next century.

A second factor is the concentration on a small scale vehicle. Rigorous limitation to light personal transport offers considerable benefits in infrastructure costs and flexibility, and also provides a system with a more human scale.

A third key feature is automation. This can allow vehicles to run more efficiently, and provides an effective means of managing traffic flows through the system while satisfying transport demands at minimum cost, and avoiding the inefficiencies of empty scheduled public transport services. Achievement of public acceptance of an automated system implies that design for safety is a critical success factor for the project as a whole.

*A Brief Description of the New System*
Figure 1 gives an impression of the system in operation. The core system is automatically controlled personal taxi vehicles, which run on a dedicated guideway network covering, in principle, the whole of the city (and potentially beyond). Passengers go to a local station, generally within 250m of any point in the city, and use a ticket machine, or other computer interface, to state their destination. A vehicle will normally be waiting at the stop. If not, one will be dispatched by the system. Present simulations demonstrate that service within 1 minute for 95% of the time. Vehicles are currently projected to have two permanent and two tip-up seats for maximum flexibility. After boarding, and confirmation of the trip, passengers are taken automatically to their destination. After alighting the vehicles are recirculated within the system. All stations are off line, so that vehicles wishing to stop at any station do not impede the passage of others. The track is passive, as opposed to the active points systems used in trains and trams. This reduces cost, allows short headways, and eliminates a possible source of failures. The ground level guideway is estimated to be cheaper than a pedestrian footway to install due to

the restriction to low weight (max 1000kg full) vehicles.[1] Overhead sections are estimated to cost less than one twelfth of present road (or rail/light rail) overhead structures. Power will be provided through pick-up from shielded electrical power supply in the guideway. Basic vehicle guidance may be likened to a kerb-guided car, the U-shape guideway form also providing psychological reassurance on the overhead sections. Later versions of the system could use alternative guidance systems which would reduce physical obstructions on ground level sections of the track. The guideway is separated from other traffic and from pedestrians. To avoid severance issues a significant part of the guideway must be overhead, although studies using Bristol as a test case have suggested that much of the track could be placed at ground level, giving important cost benefits. The guideway network is projected to be arranged at about 500m intervals. Planning considerations have therefore been a crucial element of the study from the beginning. An initial cost estimate for the full system has been completed, using estimates from Ove Arup for the infrastructure elements. Although inevitably based on preliminary assumptions, this indicates that the whole cost of the vehicles and infrastructure could be recovered at fare levels set equal to or lower than current bus fares. The vehicles which have emerged from the evaluation are car-like. Advantage can be taken of this to provide "dual mode" capability in which vehicles can also be taken off track under driver control. This provides considerable additional flexibility and attraction.

Figure 1   An Impression of the Transport System Proposed

This class of transport is known as Personal Rapid Transit (PRT). Such forms of transport have major differences from more conventional forms of public transport.

---

[1]   Regulations require pedestrian footways to accept a higher load per unit area than the present guideway design.

Any form of transport must meet stringent safety requirements. Safety is a crucial issue for PRT, not simply from the regulatory point of view, but also from the point of view of passenger acceptance. In other areas of the design such as energy and environmental impact the design targets for the new system have been set at a level which provides a major improvement over the car. One of the major issues with the current car-road system is that accident rates are unacceptable. A significant improvement over these rates is therefore a requirement of any new system.

There is also a potential problem. Fiedler and Reynolds (1995) note "Stringent regulation of new risks can increase aggregate risk levels. The reason is that new technologies, which in general tend to be safer than old technologies, can often be subjected to more stringent safety regulation simply because they are new. The result is the perpetuation of riskier (but politically more powerful) existing industries at the expense of safer new technologies that do not have a built up base of political support". Thus it is particularly important in the present system to develop a balanced view of the issues so that genuine improvements in safety can be <u>delivered</u> to the travelling public.

This paper is an introduction to some of the safety issues that need to be considered when designing a PRT system, and gives an analysis of the requirements which will underlie a formal safety case. Its aims are first, through an examination of other transport modes, to set realistic but ambitious target accident rates for PRT, and second, to identify in general terms some of the design issues associated with the safe operation of PRT.

## 2    Target Passenger Accident Rates

Jane's Transport Systems lists over a dozen fully automated systems now in use for public transport. Many new automated transport systems have not had a fatal accident. For example the Morgantown system in West Virginia USA, has now been operating for 20 years, carried 48 Million passengers, and never had a serious accident. Nevertheless it is unreasonable to expect that any major transport system will be accident free. The purpose of this section is to set acceptable accident limits based upon an analysis of those rates experienced by other modes of transportation. The Health and Safety at Work Act requires that risk should be "as low as reasonably practicable".

### 2.1 Transport Accident Rates

Accidents can range from damaged equipment with no harm to the users through to a fatal accident, with minor and serious accidents coming in between. Passenger casualty rates for all severities by transportation mode exist. As noted by Collings (1994), in Transport Statistics Great Britain 1994, these should be treated with great caution, since the differences in reporting policies and procedures between modes can result in variation in the numbers reported as seriously injured and those with minor injuries. While some discrepancy exists between the transportation modes regarding the seriously injured, this is believed to be modest, and therefore useful for this study. Fatality rates are assumed to be highly accurate. The usual convention, therefore, is to compare passenger fatality rates, and passenger casualty rates (Killed and Seriously Injured or KSI).

## Table 1. Passenger fatality rates: by mode of travel[2]

*Fatality rate[1] per 100 million*

| Mode | Passenger kilometres | Passenger journeys | Passenger hours (FAR) |
|---|---|---|---|
| Car | 0.4 | 4.5 | 15 |
| Van | 0.2 | 2.7 | 6.6 |
| TWMV | 9.7 | 100 | 300 |
| Pedal cycle | 4.3 | 12 | 60 |
| Foot | 5.3 | 5.1 | 20 |
| Bus or coach | 0.04 | 0.3 | 1 |
| Rail | 0.1 | 2.7 | 4.8 |
| Water | 0.6 | 25 | 12 |
| Air | 0.03 | 55 | 15 |

1. *1992 rates except -*
   *air, rail, water: average rates 1975 - 1992.*
   *bus or coach: average rate 1988/9 - 1992/3.*

TWMV = *two-wheeled motor vehicle*
FAR  = *Fatal Accident Rate = Number of deaths per 100 million man-hours of exposure to the risk*

For comparison Collings gives FAR figures of accidents in the home as 4, smoking as 40 and rock climbing while on the rock face as 4000.

Passenger fatality and casualty (KSI) rates are recorded can be done in three ways, by distance; by trip and by time. Depending on which way fatality and casualty (KSI) rates are measured, they can appear very safe or relatively unsafe. Table 1 shows that two wheeled motor vehicles (TWMV), however they are measured, are the most unsafe transportation mode. However, air travel, if measured by distance (passenger kilometres) is the safest transport mode, but if measured by trips (passenger journeys) it is the second unsafest. Bus or coach are either the safest or second safest transportation mode however they are measured. The reason for this

---

[2] All tables in this section are taken from the Transport Statistics Great Britain 1994. Note that there are some inconsistencies in the published figures. These can mostly be attributed to rounding errors.

obviously lies in the difference in trip time and length between the various modes, but this does complicate the choice of a suitable target statistic.

**Table 2. Passenger casualty rates (KSI): by mode of travel**

*Casualty rate[1] per 100 million*

| Mode | Passenger kilometres | Passenger journeys | Passenger hours |
|---|---|---|---|
| Car | 4.5 | 55 | 190 |
| Van | 2.4 | 30 | 75 |
| TWMV | 150 | 1600 | 4500 |
| Pedal cycle | 85 | 230 | 1200 |
| Foot | 55 | 55 | 225 |
| Bus or coach | 1.6 | 13 | 40 |
| Rail | 0.4 | 9 | 15 |
| Water | 4.6 | 170 | 90 |
| Air | 0.04 | 70 | 20 |

1.     *1992 rates except -*
*air, rail: average rates 1975 - 1992.*
*bus or coach: average rate 1988/9 - 1992/3.*
*Water: average rate 1983-1992.*

An alternative and perhaps more relevant accident statistic is the number of Killed and Seriously Injured (KSI). Table 2 above shows these KSI casualty rates by mode of travel. Again, passenger casualty rate (KSI) per 100 million passenger kilometres is best for air with a rate of 0.04. For passenger casualty rate (KSI) per 100 million passenger journeys and passenger casualty rate (KSI) per 100 million passenger hours the best rates are for rail (not bus, as they were for fatality rates) at 9 and 15 respectively.

Again, for comparison, Collings (1994) suggests that the rate for serious accidents at home is about the same as travelling by Bus or Coach, and the rate for Skiing about the same as cycling.

## 2.2 Target Rates for PRT

For a new transport system it is necessary to set safety targets which are clearly better than those achieved in current transport systems. As has been made clear from the discussion above the considerable variability in the statistics makes choice of targets complex. The difference between Fatality and KSI rates corresponds to the nature of the transport system, with higher speed systems having a higher Fatality rate than lower speed systems. For a ground based system travelling at relatively low speed it may be expected that the proportion of fatalities will be low, so that the KSI statistic will be the more important safety objective.

The base target has been set in relation to the KSI rate per hour achieved by the safest current form of transport, the train. This corresponds to a figure of 15 casualties per 100 million passenger hours. This statistic also provides an improvement of more than a factor of ten over the car. This is consistent with the general philosophy adopted throughout out the present work of achieving order of magnitude improvements over the car. The relevant targets per hour and per passenger journey can be found from the anticipated characteristics of the system, and are based on an average speed of 50kph and an average trip length of 5km. This give the figures shown in Table 3. The relationship of Fatalities to KSI Casualties for the system should correspond at least to that achieved in the car. This suggests a Fatality rate which is about one tenth of the KSI rate, as shown in Table 3.

**Table 3. Preliminary Target Accident Rates for a PRT System**

*Rate per 100 million*

| *Accident group* | *Passenger Kilometres* | | *Passenger journeys* | | *Passenger hours* | |
|---|---|---|---|---|---|---|
| *Target* | | | | | | |
| KSI | 0.3 | | 1.5 | | 15 | |
| Fatalities | 0.03 | | 0.15 | | 1.5 | |
| *Car* | | | | | | |
| KSI | 4.5 | | 55 | | 190 | |
| Fatalities | 0.4 | | 4.5 | | 15 | |
| *Best Current* | | | | | | |
| KSI | 0.04 | (Air) | 9 | (Rail) | 15 | (Rail) |
| Fatalities | 0.03 | (Air) | 0.3 | (Bus/Coach) | 1 | (Bus/Coach) |

The comparison in Table 3 shows that the target safety standards would provide a system which was as good or better than current transport by nearly all measures. The one exception is the better KSI rate per passenger km on aircraft, but it will be noted that the targets match the fatality rate per passenger km on aircraft. The projected fatality target per hour is slightly worse than the bus. It is particularly interesting to note that the Targets give accident rates per passenger journey which are significantly better than those achieved by any other form of transport.

Table 3 also gives a comparison with the figures for the car. This is perhaps the most relevant comparison since the system is designed to compete with the car, and most passengers will have been attracted from car journeys. In terms of accidents per passenger hour or per passenger km the system targets offer an improvement by a factor of ten or over in all the safety statistics.

## 2.3  Discussion:  Reality of the Target

It is naturally rather easier to set targets than to meet them.  A crucial question is whether these targets are reasonable.  The only data available on a broadly comparable system is that from the Morgantown system.  As has already been noted this has carried 48 million passengers over 20 years without a single serious accident.  This is consistent with the target figures suggested.

The target may also be compared with those in other industries.  For example the Tolerability of Risk (TOR) of death to a member of the public from a nuclear installation is set at 1 in $10^4$ per annum. The present target of 1.5 Fatalities per 100 million passenger hours corresponds to an annual rate of just over 1 in $10^4$ per passenger year, i.e. a level parallel with that set for the Nuclear Industry (on the assumption that "passenger year" is a comparable statistic).  It is generally accepted that to achieve levels "as low as reasonably practicable" a figure of 1 in $10^6$ should be aimed for.   On the other hand risk reduction is generally regarded as unreasonable if the cost is grossly disproportionate to the improvement gained.   In the present case the target of a significant (i.e. factor of ten ) improvement over the car would provide a massive improvement in transport safety delivered to the public.   Achievement of a level of 1 in $10^4$ would also correspond to the achievement of levels of software reliability at the limit of what is generally accepted as possible in the industry.  Thus it is believed that the current targets provide a good compromise, offering clear improvement without excessive cost.

It is reasonable to anticipate that a properly designed automated system would achieve low accident rates.  In most complex systems the principal cause of accidents is human failure.  For example, it is accepted that 70-80% of all aircraft accidents are due to pilot error.  As far as is known, there has never yet been a fatality to a passenger from failure of an automated system on an aircraft.  Thus the achievement of exceptionally high levels of reliability by automated systems has already been demonstrated.  However it must be assumed that an automated system is capable of failure.  The consequences depend on other aspects of the system.  In the present system the vehicles are confined to a segregated guideway.  This provides a considerable restriction to the consequences of possible problems.  Also speeds in the system are comparatively low.  Maximum speeds currently under consideration (80kph) are considerably lower than those of car, coach or rail.  It may be assumed that the vehicles will be designed to be self monitoring so that any potential mechanical failure can be identified early.  Regular maintenance and safety checks will also ensure that levels of mechanical reliability would be far higher than is the case in many privately owned vehicles.   Further, if hazardous circumstances arise then it is reasonable to anticipate that the central control system will be able to impose courses of action on other vehicles in the system to minimise any effects.  A further discussion of the details of this approach is given in Section 4 below.

In a general discussion of the safety of people movers North (1993) makes the same general points.  He notes that because PRT type systems are completely segregated from pedestrians and other traffic, completely automated and controlled to a high degree of reliability, accident rates can be anticipated to be low and that accidents due to human factors are almost entirely avoided.

## 2.4    Non-passenger Accidents on a New PRT System

PRT is a completely segregated transport system that has no direct interaction with other modes.  In this respect it is similar to a train.  The statistics used in the previous section only refer to passenger accident rates.  Other categories used for rail (and aviation), according to Evans (1994), are staff or crew and third parties. Rail accident rates also include a further important category, *trespassers,* who are people unlawfully on the railway.

An analysis of the rail accident statistics for the last 20 years, as shown in Evans (1994), suggests that staff or crew and third party fatalities account for 15-20% of all fatalities excluding trespassers (i.e. passengers account for 80-85 % of rail fatalities excluding trespassers).  Given PRT carries no crew on board, accidents involving staff or crew will be removed.

As with rail, it should be recognised that trespassing will unfortunately occur on PRT.  Elevated guideway (at 5.7m, the statutory height above roads) will segregate the guideway from other users.  Lower level elevated guideway and guideway changing from ground level to elevated will also offer a degree of segregation. Ground level guideway may have to be fenced.  This would restrict access to the guideway and prevent interference from other sources.  However, fences may not be appropriate because as well as preventing people trespassing, they prevent trespassers leaving the guideway and this may not be deemed acceptable.  This issue needs clarifying with the Railway Inspectorate.

It should be noted, according to Evans (1994), that for rail, the number of fatalities resulting from trespassing is, on average, about double passenger fatalities.  How many rail trespassers are those deliberately trying to take their own lives is unknown.

At this point it is difficult to judge the number of fatalities on PRT occurring from trespassing. PRT is anticipated not be a suicide magnet in the way rail is.  On the other hand if PRT is perceived as safe then this could attract onto the guideway people who might resist encroaching on a track which was more clearly unsafe. This does raise a very difficult area of judgement.  It is possible in principle to use a sensor suite on the vehicles and/or the guideway to give the vehicles a capability to react to hazardous situations from people trespassing on the guideway. Unfortunately provision of such protection is likely to provide an attraction to potential trespassers, since they could interfere with the system at little risk to themselves.  Thus provision of a vehicle sensor suite could lead to additional overall risk, not only to the trespassers but also to passengers in the vehicles.

An initial solution to this dilemma is to use any sensor suite as a means for deterring trespassers, for example by loud audible warnings, intense lights etc. This would require support by security management measures.

## 2.5    Implications for Casualty Statistics

The introduction of the new system would have a profound effect on casualty rates. Existing public transport systems are considerably safer than the car, but are not widely used. This is because current public transport systems which require collective transport along predetermined corridors, and provide neither the speed nor accessibility nor convenience of the car. The new system under consideration can compete effectively with the car. Examination of Bristol using established behavioural approaches for determining transport choice based on trip time/cost has suggested that up to 40% of current car users would transfer to the new system.

The safety implications of this change are highly significant. If it assumed that people will choose to take the same trips by the new system as by car, then the best single statistic to use for comparison is the KSI per passenger km for which a factor of 15 improvement is projected. These assumptions would provide a reduction of 37% in the casualties over the passenger casualties in the present system. In addition it is reasonable to project a parallel reduction in the casualty rates for other road users such a cyclists and pedestrians of the same order. Thus there is a strong case on safety grounds alone for serious consideration of the new system

However these projections are of no value unless a safe system is designed and delivered to the customer. The next section discusses some of the issues underlying safe design of the new system.

# 3 Safety Design for PRT

## 3.1    General Considerations for Safe Operation of PRT

Automatic systems will be unacceptable to the public unless they are perceived to be exceptionally safe. Thus achievement of the highest levels of safety is a critical design objective. As has already been noted the basic features of PRT design, e.g. the elimination of the driver, guideway separation, and relatively low speed are all highly positive from the safety viewpoint. However these basic features must be supported by in depth detailed design to ensure that safety targets are met and preferably exceeded.

In this section some of the basic measures that could be included in a future PRT system are examined. The current paper is only the first stage in a consideration of safety issues. These will have to be studied in considerable depth to confirm the overall benefit and, where appropriate, to evaluate cost-effectiveness of alternative approaches.

The legal framework for regulation of guided systems such as PRT is based on that for railways. However, as noted in the document setting out the fundamental top level safety principles for railways - HSE(1996), " the principles are intended to apply to ... guided transport systems only to the extent that is appropriate". Further, there are no specific regulations for such systems as there are for railways or tramways, so that safety issues must be considered from a fundamental viewpoint.

The issue of safety for the system is directly linked to the functionality of the system as a whole, and can only be fully addressed by the full range of rather formal methods, supported by a computer based description of the interactions possible on the system. Fortunately there is now reasonable experience of the design of safety critical systems so that experience from a variety of areas can be brought to bear on the problem.

The key issue is the identification and prioritisation of hazards. The safety case must be formal, and supported by the usual range of techniques such as Fault Tree Analysis, Event Tree Analysis and Failure Modes Effects and Criticality Analysis. The likelihood of each route to a potential hazard must be evaluated, and critical faults identified. Particular attention must be given to common cause faults (e.g. power failure). Much of the system will be software driven. Thus the use of established good practice for safety critical software will be a necessary part of the design.

The basic principles for failure free design are adequate design margins, elimination of design error, protection against extreme events, and protection from system degradation. The problems in any system can be classified under two main headings

1. Problems arising from failures in the system itself
2. Problems arising from outside influences

A distinction must also be made between systematic failures which are a result of inadequate design (for example software) and those which are a result of random errors or degradation of the system. In most systems the influence of human error in operation is a significant issue in the design. For the present system this issue is minimised.

The design of the system must meet a number of critical safety and operational objectives:

1. The system must have a very low failure rate.
2. The system should have in-built redundancy, so that local failure leads to zero or minimal degradation.
3. The system should be designed so that any failure has the lowest possible possibility of leading to a safety problem.

4. If the system fails passenger inconvenience should be minimised.
5. Failure should have an extremely low probability of leading to death or serious injury (see Section 2)
6. Recovery from failure should be straightforward, and lead to the minimum divergence from standard operating procedures.

Item 1 above is an essentially internal system matter. The other items on the list apply to both internally and externally induced failures. In all cases there are two basic principles: first the system must be made basically simple and safe; second, protection should be provided by system based risk reduction approaches

An important issue is that of intrinsic safety. It is common practice in railway systems to adopt a design which will shut down in the case of any malfunction. This approach is not practicable in a more complex system. For example aircraft, which do not follow this process, have achieved a good safety record. A key principle here is fault tolerance. In the present case many sub-elements and/or sub-systems will be designed to operate autonomously. It will be necessary to develop robust procedures for identifying problems which can reasonably be overcome, and distinguishing those which do require immediate priority action, overriding autonomy of sub-system elements. Formal system description methods will assist this process.

## 3.2 PRT System Features for Safe Operation

The following features of the system are expected to contribute towards the achievement of high safety levels. Note: the order given below corresponds to the order in which they are subsequently discussed, not to an order of priority.
- fault tolerant design
- intelligent vehicle
- hierarchical system design,
- separate merge logic
- collision avoidance and protection suite
- health and status monitoring/reporting from the vehicle
- video surveillance of stations
- voice communication with the vehicles
- smoke/fire detection within the vehicles
- fully developed automatic and manual safety procedures
- vehicle problems normally dealt with off guideway
- run flat tyres
- back-up battery power supply
- network re-routing in the case of failures

A basic principle of the design is fault tolerance i.e. single failures at any part of the system do not interfere with normal operations. A crucial element of the control system design philosophy is the balance of intelligence on the vehicle and infrastructure. It seems highly desirable to put as much intelligence as possible in the vehicle. The infrastructure has to deal with combinatorial problems, so that it is important that the basic decisions being made are kept to the minimum level. In contrast, the vehicle, in principle, has a straightforward task. Furthermore any computing on board the vehicle will be mass produced and can expect to have a low cost level.

The key function of the central computer is to act as a scheduler. It will assign routes and timings to individual vehicles. Once the vehicle is advised of its route it will proceed autonomously. The central computer is then only required to take further action in the event of a major fault which invalidates the route advised. All vehicles will also need to report back position to central control so that proper fulfilment of the route instructions can be confirmed. Thus an important area both for safety and for operational reliability is the central control system. The challenge is that the control system should never break down. Hence it needs at least one back-up power supply to continue functioning safely. Back-up computers are essential for the central system since any failure here affects the whole system. Ultra safe systems in aircraft have featured triplicated approaches. It is not clear how much genuine benefit in safety has been achieved by this approach. It is unusual for such systems to meet the same operational requirement via fully dissimilar internal functionality, so that the likelihood of common mode failures is high. Further, many systems designed with multiple redundancy have suffered from failures at non-redundant links. There have also been examples where the duplication or triplication of systems has been the cause of failures. The cost of providing fully duplicated and functionally dissimilar control systems is very high and there is only limited evidence of proportionate benefits. Where there is an issue of safety of life the highest priority must be given to prevention. However, in the present case, many of the problems that might occur are more likely to result in a minor incident rather than serious accidents. Thus it is suggested that the most appropriate approach to system safety will be to arrange for fully duplicated computers, power supply etc, with a monitoring circuit checking reasonableness of the output. In the case of outputs which go outside limits then the computer system would be switched. If the fault remained then the system would be shut down. Overall system shut down is simple since all vehicles can coast to a halt at constant separations. Restarting the operation will be rather more complex, and is very likely to involve human intervention. Operating procedures for this process will need to be developed.

The basic design of the system, based on a fairly dense network with off-line stations, is such as to minimise the consequences of single failures, whether of a vehicle or of part of the infrastructure. In the event of a major vehicle or local infrastructure fault then only the local branch needs to be closed, and the network will automatically re-route vehicles around the fault.

The computer on board the vehicle is responsible for ensuring that the vehicle meets the route imposed by the central computer. Basic inputs to this will be dead reckoning from known vehicle wheel rotation, calibrated by interrogation of way points around the track. It is less clear if the vehicle computer also requires duplication. Failure of computing in an individual vehicle only puts a single vehicle at risk. The key requirement here is to incorporate procedures which permit the vehicle to overcome this failure without risk to the occupants. This can be accomplished by arranging a separate system to provide for removal of the vehicle to the nearest station on the event of any problem.

One of the most significant potential safety issues on the system as a whole is the merging of two vehicles onto the same guideway. Faults in this process do have the potential of causing accidents. It seems that it would be wise to devote a special controller to the merging process with merging logic supported by duplicated checking and reporting procedures.

Protection of the merge process means that the most obvious  hazard is the two vehicles colliding on the track. The optimum solution to protect the occupants in these circumstances will include a combination of several elements. In nearly all circumstances vehicle problems will cause a deceleration rather than an instant stop. Thus it is very feasible to have some form of advance warning to a following vehicle that the vehicle in front has become a potential hazard. This would allow the following vehicle to brake to a halt. Thus collisions between two vehicles would require simultaneous failure of different systems in the front and following vehicle. This would be a very unlikely occurrence.

The possibility of the appearance of a major hazard on the track , for example a fallen tree would require other approaches.  Standard approaches to crash management in cars, such as air bags, could also be incorporated. However it will be necessary to undertake a systematic hazard study to ensure that the safety package finally used is properly balanced between the various hazards.

A basic feature of the vehicle design will be a health monitoring system. No vehicle will be allowed on the system unless it has passed a check that all key operating parameters are within limits. These will also be monitored en-route. The vehicle will also report status to the central system control which will check that all status reports are consistent with expectations. Should parameters/status go outside limits then appropriate actions will be taken. These could include, in order of severity, allowing the vehicle to continue in service, progressing to the destination station and then being removed from service, and progressing to the next available stop for passenger egress. Minor degradations will involve the judgement of the operational controller. More severe degradations will invoke an automatic response. The only circumstances currently foreseen in which the system would cause a vehicle to stop on the guideway will be if the vehicle in front has come to a halt. As will be discussed in more depth below it is believed that in all other circumstances the safest option will be to proceed along the guideway to a suitable stop.

It is normal practice for transport systems with automatically controlled vehicles to have all stations equipped with closed circuit television cameras (CCTV) allowing central control to view the entire platform area. This feature is important not only for safety, but also for security of passengers. It is likely that city centre stations, or stations handling large levels of demand, would be manned.

It is also normal in automatic systems for a communication link to be supplied to each vehicle, and at each station, giving travellers immediate two-way communication with central control. Experience from the Morgantown system suggests that this is an important element in providing passenger reassurance when needed. The communication system can be used for provision of information from central control to travellers at stations or in vehicles during normal operational circumstances, but will be of particular significance in emergencies.

Risk of fire in the system is very low. Electrical power eliminates the carriage of inflammable fluids in the vehicles. The occurrence of electrical shorting is a possibility but this can be reduced to very low levels by established electrical design methods. The vehicle on-board health monitoring system will provide forewarning of any degradation which might lead to electrical faults. The vehicle will be designed from fireproof materials. The risk of externally induced fire is very small. The crucial area of the vehicle is the passenger compartment. Each vehicle could be provided with smoke and/or heat detectors relaying to central control. This implies that smoking will be forbidden on the vehicles. If smoke is detected then vehicles would immediately be re-routed to the nearest fire management point. In general this would be the nearest station.

A full set of emergency procedures will be developed to deal with all foreseeable problems. The preferred plan for any emergency is to get the vehicle(s) automatically to a stop where the passengers can alight and exit the system. Since stations are only 500m i.e. 50 seconds apart it appears far more practical to isolate any problems at a station or other suitable location rather than to attempt to deal with problems in transit. Nevertheless, it must be assumed that problems can occur where the vehicle does come to a stop on the guideway. Although these can be anticipated to be a very small proportion of overall failures they are an issue which must be addressed. It is normal practice in equivalent systems e.g. monorail, cablecar etc to require the passengers to remain in the cab if failures occur. It is believed that this will provide the safest option in the present case. Evacuation of passengers is likely to offer far higher risk than dealing with them in situ, especially if they are elderly or disabled.

The solution adopted in other equivalent circumstances is to arrange for the failed vehicle, and any other vehicles stopped as a consequence, to be pushed or pulled to the nearest station. To aid this process it is probably desirable that vehicles are able to reverse in an emergency. There may be a few occasions where the option of moving the vehicle directly is unavailable or impractical. If the vehicle is at ground level then passenger egress is straightforward. However if the vehicle is on an overhead section passenger egress from the vehicle is likely to incur high risk. Also, more severe accidents may incur damage to the guideway, so that it would again be safer for passengers to remain in the vehicle until assistance arrives. The solution to this in other transport systems, which is also recommended in the present case, has been to arrange for special equipment, together with specially trained staff, to be available. These are brought to the stranded vehicle to undertake the rescue operation. Passengers need to be made aware of these procedures, so that provision of a two-way communication link to the vehicle is an important element of the approach.

The only alternative to this process would be the provision of walkways. These would significantly increase the cost and visual intrusion of the system, almost certainly to the point where the system would not be viable, and provide little improvement to safety. Thus on those occasions when evacuation from the vehicle is necessary, it seems sensible to adopt an in-vehicle recovery process, either on track, or off track via an emergency recovery vehicle, as the normal standard.

The current configuration for the proposed PRT system for Bristol is of a rubber wheeled vehicle running on asphalt or steel. Given the highly managed environment of the proposed system this should cause fewer problems than conventional cars running on asphalt. However, there is always the possibility of a tyre blow-out or puncture. This would affect the steering and electricity pick up from the power rail. A number of solutions exist. The most obvious would be to run solid rubber tyres. This would require an alternative approach to the vehicle suspension. The other would be to use a run-flat tyre. A review of tyre technology would identify the best solution. In the event of a problem with the tyre, the vehicle would be directed to the nearest station or to the depot.

Loss of power to the vehicle needs to be considered. The current concept for power supply is electric power from a power rail, supported by a modest on-board battery. This means that where there is a loss of power supply from the power rail, the vehicle should be able to complete its trip by running from the battery. A complete internal power failure would mean the vehicle would come to a stop on the track. This probability will be minimised by redundancy studies during the design. In this situation the vehicle would have to be shunted or removed from the line.

A vehicle stopping on the track presents a number of safety problems that need to be overcome. As has been discussed it is most unlikely that an vehicle would stop dead on the track. In the event of power loss for example the vehicle would coast to a halt and allow vehicles behind to take action either to brake or re-route. Power loss would provide a status report to central control. Once the control system knows there is a problem, trips would be re-routed to avoid the stretch of guideway with the blockage. Those vehicles which are behind the stopped vehicle and cannot be re-routed also need to be stopped, and moved to a location where they could be evacuated. This reinforces the advisability of equipping the vehicles with a reversing capability.

The loss of a link could cause capacity problems on the system, especially if it is an important connection. This problem reduces as the network grows because there is an increase in the number of alternative routes that a vehicle can take. Hence, the loss of certain links on a small scale demonstration system could constitute a major system failure. A reduction in system capacity would prevent some trips from beginning. This information would need to be conveyed to all stations.

## 3.3 Other Operational Considerations

There are several further related operations issues. Vandalism and attacks on travellers is one such problem, not unique to PRT. The most suitable approach to this is via security management. It is believed that the present systems should generally be less prone to these problems than other forms of transport. PRT provides a near instant service which means little time is spent waiting at the station, the journeys are quick, and passengers chose their companions for the journey. Also, it is possible to provide a high level of security at the stations, and possibly in the vehicles, in the form of CCTV and two-way communication. These should create a safe and secure system for both passengers and for staff. City centre stations are likely to be manned and both increase safety and reduce vandalism.

Consideration could be given to mounting CCTV within the vehicles themselves, but this would be principally a security issue. A PRT system covering the whole of Bristol which consisted of several hundred stations and possibly several thousand vehicles constitutes a major surveillance task, and some degree of automation/intelligence would have to be considered. This is not seen as an essential element of the design at present.

The most effective deterrent, to both crime and vandalism, is a high probability of being caught. Video surveillance would need to be supported by effective security management. Problems such as graffiti would be addressed by standard approaches. These include the selection of surfaces for both stations and vehicles which do not encourage graffiti, and which are easy to clean in the event of difficulties. Both crime and vandalism have a strong element of imitation, so that elimination of the problem at the earliest possible stage would probably prevent its development.

Normal operation needs to be guaranteed during all weather conditions, including very hot weather, rain, snow and ice. For ice and snow, guideway heating may need to be employed. Ice on the guideway will affect the performance of the vehicles and if guideway heating is not employed, the infrastructure design will need to take account of reduced vehicle performance to ensure continued safe operation. A number of snow/ object clearing utility vehicles will need to be employed at different times to ensure continued safe operation of the PRT system. Leaves and rubbish would also have to be kept clear of the guideway. Weather should only prevent operation in extreme circumstances, for example, severe gales or blizzards. It is suggested that safe operation should be guaranteed up to at least 1 in 5 year type storms and preferably 1 in 10 year type storms. However it is believed that passengers would be likely to accept reduced levels of operation, e.g. reduced speeds, under difficult conditions. The best approach to this will come from operating experience.

# 4    Conclusions

This paper has provided a general introduction to the issues associated with safety of Personal Rapid Transit (PRT). The safety targets proposed, based on 15 Killed or Seriously Injured per 100 million passenger hours, meet or exceed those obtained in other forms of surface transport. In particular the target rates give an improvement of more than a factor of ten on those currently achieved by the car. It appears that such targets do represent a reasonable objective for a PRT system. Such levels would provide a major reduction in traffic casualties in areas in which the new system was introduced. A discussion has been presented of approaches to ensure the safe operation of PRT. Further study is required to assess these in depth, prior to the development of a formal safety case.

## Acknowledgements

Thanks are due to Gordon Hughes, John May, and John Napier of the Safety Systems Research Centre for valuable advice during the preparation of this paper.

## References


Collings, H. (1994), "Comparative Accident Rates for Passengers by Mode of Transport - A Re-visit" Transport Statistics Great Britain 1994, HMSO, London.
Evans, A.W. (1994), "Multiple-Fatality Transport Accidents: 1946-1992" Transport Statistics Great Britain 1994, HMSO, London.
Fiedler A. and Reynolds, G.L (1995) "Legal Problems of Nano-Technology - An Overview" Southern California Interdisciplinary Law Journal; Vol 3 No2 pp593-629
Health and Safety Executive (1996) "Railway Safety Principles and Guidance" ISBN 0-7176-0712-7
North, B.H. (1993), "A Review of People Mover Systems and their Potential Role in Cities" Proc. Instn Civ. Engrs Transp. **100**, May, p95-100.

# Safe Systems Architectures for Autonomous Robots

I. Sommerville, D. Seward, R. Morrey and S. Quayle
School of Engineering, Computing and Mathematical Sciences
Lancaster University, LA1 4YR, UK
E-mail: is@comp.lanc.ac.uk

**Abstract**   This paper describes work we have carried out in the development of a control system for an autonomous robot excavator. We describe some specific problems which complicate the development of safe systems for autonomous vehicles. We then go on to discuss how we developed an understanding of the hazards of vehicle operation and describe the architecture of the excavator control system. In this architecture, all system actions are moderated by a safety management sub-system. This system checks all movements against a set of potentially hazardous situations and information from independent sensor systems. Potentially dangerous actions are curtailed or corrected by the safety manager. We also discuss the trade-offs which must be made between safety and economic viability in this type of machine and  suggest some areas where we believe that further research is necessary.

# 1      Introduction

Over the past three years, a team in the Computing and Engineering departments at Lancaster University have been involved in developing an autonomous robotic excavator. LUCIE  (Lancaster University Computerised Intelligent Excavator) is based on a commercial manual hydraulic excavator, but has an on-board computer system in place of a driver to control the hydraulics and therefore the machine. It is being developed with one particular task in mind: the digging of foundation trenches on a building site. The ultimate aim is to develop a machine which will be able to accept a program of trench locations and dimensions, traverse a building site and dig a series of trenches meeting these specifications. It should do this without the need for a driver or any other kind of human intervention.

The digger is a JCB 801.4 mini excavator (see Figure 1). All of the excavator's movements are hydraulically driven and it is possible to clearly define the working envelope at the limits of reach.  The objectives of our work were to investigate the safety problems which are inherent in this type of vehicle and to design and prototype  a control system architecture which allowed for safe, unmanned operation. This paper is mostly concerned with the control system architecture and the rationale for that architecture. However, we also discuss some of the particular problems of hazard analysis which we faced in designing this architecture.  A more general description of the system and the associated research project has been given in previous papers (Seward *et al.*, 1995; Seward *et al.*, 1996; Seward and Quayle, 1996).

Our starting point for the work was an operational prototype of the excavator which had taken no account of safety considerations. The automation of this existing prototype was based on technology which could not readily be scaled up to handle the new features and the safety checking which we felt were essential. We therefore had to completely re-design and re-implement the vehicle control system. We also had to plan the situation of environmental sensors on the vehicle and interface these to the control system. The need to re-design the system was a positive advantage as we were not faced with the very difficult problem of grafting safety checking onto an existing system.

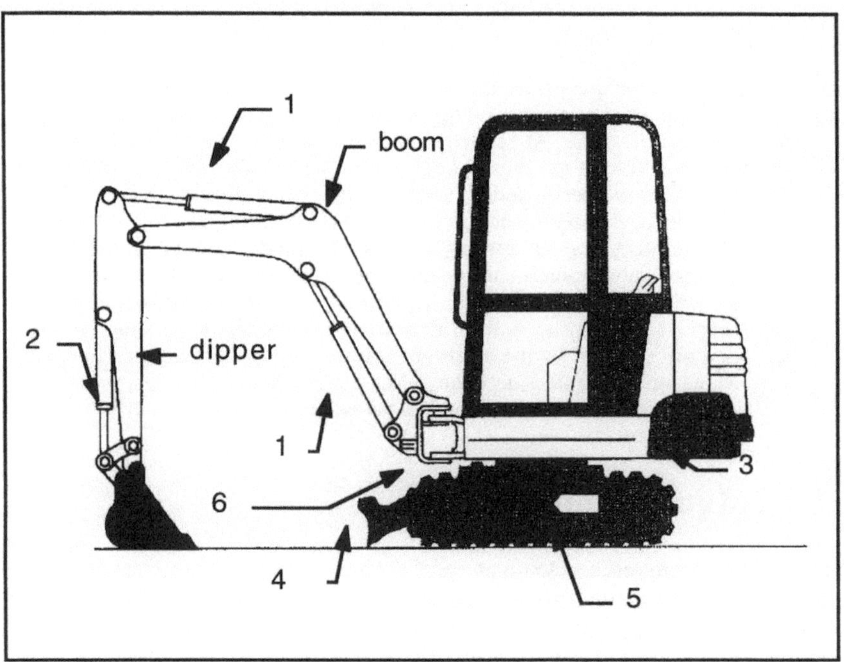


**Figure 1 - Cross-section of LUCIE**

# 2    Safety Problems

Autonomous vehicles and robots offer significant safety advantages as they can work effectively in environments which are potentially hazardous to people. However, they present their own safety problems which have not been widely explored by the safety-critical systems community. We were aware of some of these problems when we started the work on LUCIE - others have emerged during this research. These problems are:

- It is even harder than normal to define safety standards for these machines - the level of accidents which is normal in manned vehicles such as cars is probably socially unacceptable in unmanned systems.

- Current techniques of safety analysis, such as fault trees and FMEA, do not work well with these systems.

- Robots, by definition, are user programmable. We must allow for these programs being incorrect.

- The systems are designed to operate in a wide variety of relatively unstructured environments - it is practically impossible to define all of these environments in advance.

- To operate in these environments, new technologies such as neural nets may have to be used - we do not understand how to assess the safety of systems built using these technologies.

*Safety standards*

The establishment of safety standards is always influenced by social and political factors. This is particularly true where the system works in public places, where an automated system replaces a previously manned system and where it is obvious to an observer that the system has potential for harm. Unmanned vehicles are particularly emotive and we simply do not know the level of safety which will be socially acceptable for automated excavators. The driver of these vehicles is most likely to be injured in any accident so removing the driver should reduce the overall risk; nevertheless, we suspect that the level of safety which will be demanded by the public will actually be significantly greater than that expected of current manned systems.

Deciding on the acceptable level of safety is complicated by the fact that previous safety data is not publicly available. We therefore concluded that the only feasible approach was to reduce incidence of an accident until it became ALARP (As low as reasonably practicable). We are not entirely happy with this, however, and we believe that further research is required to develop a better understanding of acceptable and realistic safety criteria.

*Safety analysis*

After a number of experiments, we came to the conclusion that analysis techniques such as fault-tree analysis and failure mode effect analysis were of little relevance for the type of safety analysis which we needed. These techniques can be effective for discovering the relationship between design faults and safety-related system failures. However, they do not adequately address the earlier stages of the life cycle where there is a need to develop an understanding of what is meant by safe operation and to establish limits on the operation of the system. We found that existing techniques could not cope with complex situations which arise in unstructured environments as these situations could not readily be represented as combinations of discrete events. We return to this topic later in the paper.

*User-programmability*

Safety-critical software is normally developed by skilled and experienced staff and is rigorously checked as part of the system validation process. This is certainly possible for some of the control software in robots and unmanned vehicles. However, the diversity of tasks faced by these systems and the environments in which they operate means that some facilities must be user-programmable. We do not think it possible to restrict this to actions which are inherently safe.

We must therefore design these systems on the assumption that they will be programmed by end-users and on the assumption that any end-user programming cannot be validated or trusted. There is obviously a difficult trade-off here. The more we can restrict end-user programming, the easier it is to provide safeguards against error. However, restricting this type of programming may reduce the range of environments in which the system can operate and hence the marketability of the system.

*Unstructured environments*
Our particular problem domain differs from most other problem domains on which safety analyses have previously been carried out. These domains have almost always involved structured environments. That is they are concerned with systems which control man-made environments, particularly chemical production plants  and nuclear installations. Even in the aerospace industry, control systems are concerned with the reliable provision of  flight controls to the pilot and therefore interface not to the outside world but to the internal mechanics of the aircraft.

In our case however, the system interacts directly with a natural environment: a building site. These sites can vary radically as far as their situation is concerned (e.g. from tropical conditions with extremes of heat and rainfall through to winter operation in snow and sub-zero temperatures) and in the physical characteristics of the site (ground conditions, site geology, etc.).

During normal operation, the system must establish the state of its environment, the locations of obstacles, its own location etc., and take actions to change that environment, move to a different location, dig a bucketful of earth etc. Thus our problem domain is unusual in the field of safety analysis,  because it involves direct interaction between the system and an *unstructured environment*. It is practically impossible to anticipate the range of operational conditions and to program the machine to cope with all of these conditions.  However, it is commercially unacceptable to place strict limits on the machine's operating environment.  Any environmental assumptions which are made in the safety analysis and validation of the system, therefore, must be very generic.

*Unverified technologies*
Although not implemented , we have investigated the possibility of using neural nets to detect underground objects which might be damaged by normal excavator operation.  Neural net systems and other AI technologies may be the only way in which we can provide the system with the capability to operate in unknown and unstructured environments. Yet, the safety implications of these technologies and their interactions with other parts of the system are extremely difficult to assess. Neural net programming is clearly like user-programmability in that we must assume that errors will be made; AI technologies, because of their complex search strategies are not susceptible to the currently available techniques of hazard analysis, formal specification or safety validation.

# 3    System Safety Management

A consequence of the problems discussed in the previous section is that it is impractical to develop the system control software using the 'accepted' techniques of safe software development. That is, we cannot produce a detailed and complete specification, analyse this specification, develop the software using best practice and validate against its specification. Rather, we need some other approach which will allow us to maintain safety in the presence of potentially unsafe system components.

The solution which we have adopted is based around the concept of a system safety manager as an autonomous processor ultimately responsible for all aspects of system safety. This can be seen as a development of the concept of a software safety executive (Leveson 1985). The safety manager is an independent process which monitors all system actions and has access to all sensor information. This allows it to make judgements about the potential safety of any particular action. If any particular action is judged as likely to result in an unacceptable risk to the system or its environment, the safety manager can pre-empt that action and either modify it in some way or issue an alternative, higher-priority command which causes the system to make a transition to a safe state. This concept is illustrated in Figure 2.

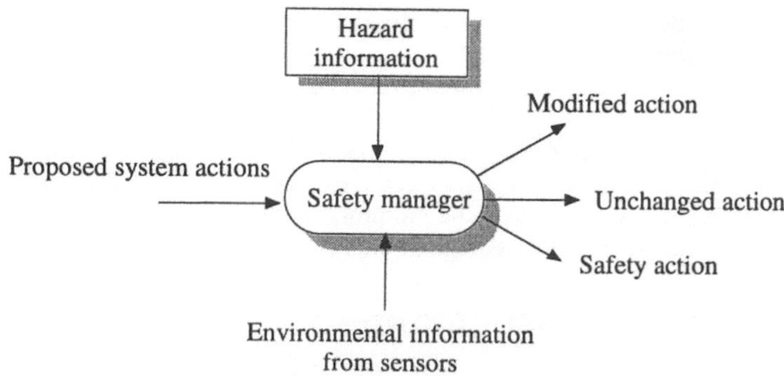


**Figure 2 A system safety manager**

The centralised management of safety has several clear advantages over approaches where responsibility for safety is spread across the whole system:

1.  The design of the other modules can be simplified as they need only focus on their principal tasks rather than worrying about safety considerations. This reduces both development and validation costs.

2.  Modules are free to use non-deterministic techniques such as fuzzy logic and artificial neural networks to carry out their computations. It is only the end-result of the computations which are important.

3.  Incorrectness (e.g. from user programming) can be tolerated. The safety manager should be able to ensure the system is returned to a safe state in the event of failure or malfunction of one of the other modules.

4.  If one module can guarantee safety whatever the status of the rest of the system, then that module only will be safety-critical. Limited resources may be focused on that module to ensure its safety and integrity.

5.  System maintenance is simplified. Modules can be replaced as new technology becomes available without extensive system re-validation.

6.  If unanticipated hazards are discovered during system operation, provision for these hazards can be made by re-programming the safety manager. Other parts of the control system may not need to be modified.

Of course, this centralised approach to safety management is not appropriate for all classes of system. There is an inevitable overhead in moderating all system actions centrally. The approach is probably only viable where there are a relatively small number of generic hazards rather than hazards which are specific to particular sub-systems. It also requires that there are known safe states of the system which may be reached from any other system state.

We believe that these conditions probably apply to most autonomous vehicles and robots. The hazards for these systems are mostly collision hazards of some kind or hazards which threaten the operation of the system itself. While there are other hazards, such as fire hazards, these are not a consequence of automated operation. They exist in manned vehicles. Safeguards have already been devised and these should not be the responsibility of the control system except insofar as the system safe state must take them into account. In general, we can ensure safety in these types of system by reducing their operating envelope and, where necessary, cutting power to the system.

To implement a safety manager we must first develop a detailed understanding of the hazards that the system might face in normal operation, design a sensor system to provide information which will allow the conditions which result in these hazards to be detected and design a system architecture which allows for all actions which are potentially dangerous to be monitored. We discuss the problems of hazard understanding and architectural design in the following sections.

# 4    Hazard Understanding

The first stage of our work was to carry out a detailed safety analysis of the system where we tried to identify the specific hazards which the system was likely to face and the probable causes of these hazards. Although we obviously had an intuitive knowledge of the hazards of excavator use, we did not have detailed knowledge of when these were most likely to arise and what actions were necessary to counter these hazards.

A hazard definition must necessarily be the definition of a whole *class* of real situations i.e. all those situations which the definition does not exclude. The difficulty is particularly acute in our problem domain, where definitions of dangerous situations must include the state of a complex unstructured environment. A hazard analysis technique must elaborate on the means of hazard definition in two ways. Firstly, it must define *how large* the classes delimited by each hazard definition should be. In general terms the classes should be small enough so that all their members have useful similarities. At the same time, the classes should not be so small that hazardous states are artificially excluded because they do not fit into very narrow and rigid descriptions. Secondly the analysis technique should define *where* the boundaries should be drawn between each hazard class.

In our analysis we started off by defining hazards in the most generic way possible (i.e. used large class sizes), to avoid making any initial assumptions about system and environment or any interactions between the two, and to make sure no hazardous situations were omitted from the analysis before we started. We eventually classified the hazards as follows:

1.  Collision with an object on the surface

2.  Collision with an underground object

3.  Toppling of the excavator

Certainly there is an element of 'physical effect' in the use of the terms 'collision' and 'toppling'; and this seems one of the best ways of creating hazard definitions which are all-inclusive. However the separation between collisions with surface and underground objects also embodies some knowledge about 'system action'.

Once we had identified these broad classes of hazard, we attempted to apply Fault Tree Analysis (FTA) to these hazards. We found that this approach was practically useless in developing a deeper understanding of the hazards. The problems we faced are due to two factors:

- Fault trees were devised to establish sets of design faults which in combination could give rise to a particular hazard. Because we were carrying out our analysis at an early stage in the development process and we were interested in hazards that could arise during normal operation, we found that it wasn't possible to relate the identified hazards to faults. The hazards were mostly a consequence of the interactions between the machine and its environment, not excavator system failures.

- Most of the systems on which FTA has previously been used seem to have been control systems which work in structured environments. Unstructured environments give rise to very complex situations which are impossible to represent as conjunctions and disjunctions of discrete events. Indeed FTA's lack of suitability for representing dynamic events, like toppling, has already been recognised (Hope 1993). This limitation also applies to Event Tree Analysis, and all other technique which represent situations as combinations of discrete events.

Failure Mode Effect Analysis (FMEA) was considered as a possible hazard analysis technique but was seen largely as a bottom-up approach. It is best suited to back analysis of existing designs, and consequently not suited to early use in the development process for innovative systems

*Physical hazard analysis*
After this false start, we decided that it was most appropriate to try to understand the hazards in terms of the basis physics of machine operation. For example, for the toppling hazard, we expressed, in mathematical terms, why toppling occurs. For an object to topple, its centre of gravity must act outside the area of its contact with the ground in at least one vertical plane. Three quantities directly determine whether or not an object will topple in a particular plane: the location of its centre of gravity in that plane, the length of its base in that plane, and the angle of its tilt in that plane.

Having established this theoretical basis, we could go on to suggest factors which might affect these three quantities. The position of the centre of gravity is affected by the orientation of LUCIE's moving parts, principally the boom arm, and the mass of any load in the bucket. The length of the base in the vertical plane obviously varies because the base is not circular but rectangular; the angle of tilt is influenced firstly by the slope of the ground in any one plane and secondly by tilting off the ground as the result of an external rotational force. Wind pressure on the machine can also contribute to toppling.

With these factors in mind, we returned to the description of LUCIE's basic functionality, in order to establish:

- At which points during operation (if any) there is danger of toppling

- How the hazard might best be avoided, or failing that detected and an accident avoided.

We determined that LUCIE's most unstable position — the position when LUCIE's centre of gravity is closest to the edge of the base in any one vertical plane — is taken up during the action 'Release Earth'. LUCIE takes up a position with the cab slewed at 90 degrees, so the boom arm is in the vertical plane where the base is narrowest; the boom arm is extended and the bucket is full, shifting the centre of gravity far over to one side.

This understanding helped us to decide that a tilt sensor was necessary to measure the angle of tilt and that the values returned by this sensor had to be interpreted depending on the particular machine operation. The safety manager had to have direct access to readings from the tilt sensor and the current machine operation. The safe or unsafe angles of tilt depended on the operation of the machine and its context of use.

We have discussed the problems of hazard understanding in more detail elsewhere (Seward *et al.*, 1994; Margrave and Seward, 1994).

# 5   Economic viability

In this section we would like first of all to discuss something which pervaded our whole analysis: the necessary trade-off between safety and viability. The inherent difficulties of developing safe control systems are compounded when the systems are not one-off systems but are embedded in manufactured products where hundreds or thousands of these products might be produced. In these cases, it is essential to use mass-produced components, minimise the number of components and component interconnections and avoid costly manufacturing processes. Even relatively small additional costs can determine whether or not a product is a commercial success so safety assurance through 'over-engineering' is not a viable approach.

The overall aim of our work was to come up with a design for an automated excavator which would be both viable and operate safely. Our raw materials for this were:

1.  A reasonable idea of the algorithms the excavator would use to carry out its basic task

2.  A knowledge of the sensor hardware which would be available.

3.  Some knowledge of the environment in which the excavator must work.

What we found when investigating each of the three hazard classes mentioned earlier, was that, when trying to come up with a scenario in which available sensors could be deployed in order to avoid the occurrence of an accident, there was a trade-off to be managed. This was founded on the fact that there are two ways of ensuring safe operation:

(1)  Add extra functionality to  handle a dangerous situation which may occur.

(2)  Place limitations on the environment, or nature of the engineered system's interface to that environment, in order to ensure that  a dangerous situation does not occur.

If (1) is not scientifically possible or is impractical given the resources of the developers, then (2) must be invoked. However (2) reduces the usefulness of the system, and we had to take care that viability was maintained. Thus, when we had to make decisions between (1) and (2), we found it useful to follow the following procedure.

- For each hazard we began by assuming an environment with no restrictions, and attempted to deploy available resources to prevent the occurrence of any accident arising from the hazard given that environment.

- If safety could not be guaranteed, then we added the least possible restriction to the environment and tried again, and so on until safe operation could be guaranteed.

- At this point the restrictions were examined to see if they precluded viable operation.

- If they did, then recommendations were made for extra sensing capabilities.

In short, we made placing restrictions on the environment a last resort, in order to have the greatest possible functionality while still maintaining safety.

This type of trade-off situation occurred during the analysis of every hazard. For the toppling hazard there was a trade-off between allowing operation on land that was not completely level and making sure toppling was not a serious danger. For the hazard 'Colliding with a surface object', there was a trade-off between allowing other vehicles and people on the building site and making sure a collision could not occur. For the hazard 'Colliding with an underground object' there was a trade-off between allowing operation over land where the existence and location of underground utilities is unknown, and ensuring that LUCIE cannot cause damage to any of those utilities. These trade-off decisions gave us another reason for making the initial hazard definitions so generic. Part of the aim of our hazard analysis became the *specification* of the environment in which LUCIE could work safely. If we had described hazards in more detail initially, we would have inevitably made assumptions about the environment which would have distorted and restricted the analysis.

# 6  Control System Architecture

In designing a control system architecture for LUCIE, a number of requirements had to be taken into account:

1. There was a need for low-level control of the mechanics and hydraulics of the vehicle. The particular systems which had to be controlled were the boom arm, the bucket and the system cab and the independent caterpillar tracks for vehicle movement.

2. There was also a need for a higher-level machine control system which took input from a specification of the digging tasks that the machine should perform and which decided on the specific actions required to meet that specification. Ideally, the specification should be expressed directly as site designs produced using some CAD package. We anticipate that this will have its own set of problems but we did not investigate these in this project.

3. There was a need for centralised safety management as discussed above.

To provide information about the system's environment, we had available a number of sensors. These include a tilt sensor which can measure the angle of the machine, a collision sensor to detect objects around the machine and a position sensor to locate the machine on a site. These clearly, must be interfaced to the rest of the system.

Our original architectural design was based on a replicated communication channel between the activities manager and the safety manager and the sensor systems. The activities manager interacted directly with the safety manager to explicitly request permission to carry out an action. This is illustrated in Figure 3.

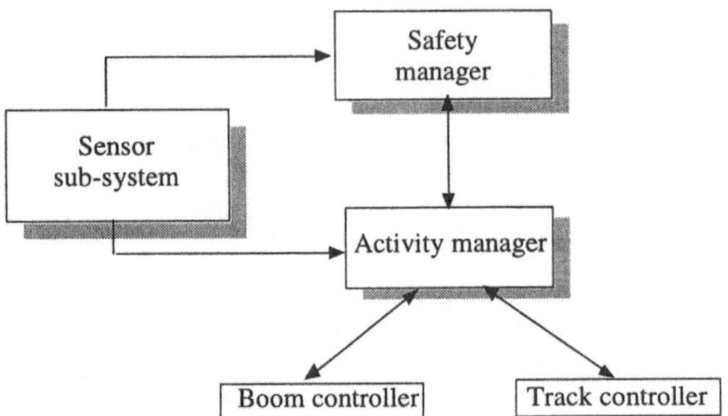


**Figure 3 System architecture with replicated communication channels**

This approach appears to be realisable but it is expensive as it requires specialised hardware to be built which supports duplicate channels from the sensors to the rest of the system. We also discovered that it did not cope adequately with a situation where the machine was in a safe state when the command was issued but a transition was made to an unsafe state during the execution of the command. This situation can arise because commands from the activity manager are high-level commands (e.g. move to location X). There was therefore a need for the safety manager to be able to take control of the low-level processes directly to avert potentially dangerous conditions.

While we were considering these problems, it became clear that a communication standard for the automotive industry (CANBUS) was emerging and off-the-shelf CANBUS components were becoming available. We therefore decided that we should develop the system architecture around the CANBUS rather than using the special-purpose communication systems which we had originally envisaged.

The system architecture which we have finally developed is shown in Figure 4. The basic responsibilities of each process in Figure 4 are as follows:

1. *Boom controller* This interacts directly with the machine's hydraulics system, opening and closing valves to move the boom arm, bucket, and rotate the cab. It executes these movements in response to low level instructions from the Activities Manager.

2. *Track controller* This operates at the same level as the boom controller, but controls the movement of the independent caterpillar tracks, again in response to low level messages from the Activities Manager.

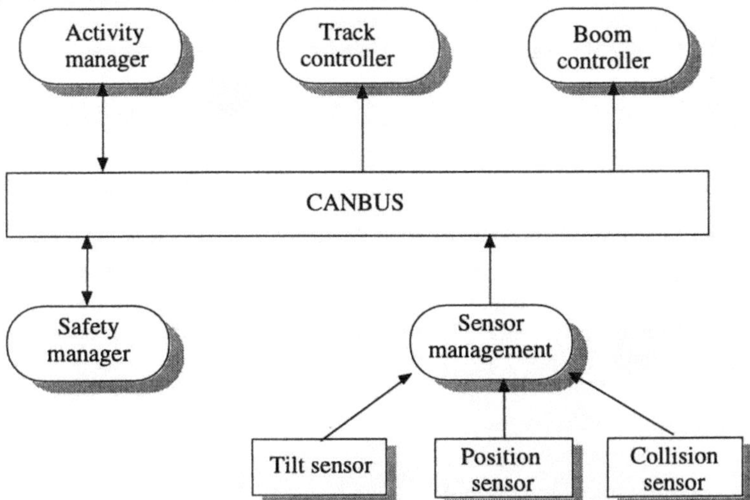

**Figure 4 LUCIE's control system architecture**

3. *Activities manager* This interfaces to the user, taking as input a program of locations and dimensions of trenches to be dug, and issues commands to the boom and track controllers to achieve the digging of those trenches.

4. *Safety manager* The safety manager captures all commands on the CANBUS and keeps track of the machines actions. On the basis of sensor information, it can take over control of the system in the event of a potentially unsafe situation arising.

5. *Sensor manager* The sensors connected to the system include a tilt sensor to measure the angle of inclination of the machine, a sensor which detects digging resistance and a movement sensor which can detect the presence of objects near the excavator.

The CANBUS is a broadcast bus so all information on the bus is available to all processes. This has forced us into a slight change to our conceptual model in that it is no longer necessary for the safety manager to moderate all commands which are issued by the activity manager. Rather, the safety manager is aware of commands issued by the activity manager and thus knows what the machine is doing. It continually monitors all sensors and, if a potential hazard is detected, it issues a further command which allows it to take over control of the machine. Once this command has been issued, the boom controller and the track controller ignore any further commands from the activity manager and only respond to commands from the safety manager.

Once the safety manager has taken control of the machine, it can do one of two things:

1. *Nothing at all* It monitors the environment to see if there is still a potential hazard. In many cases, the hazard will clear itself - for example, if there is a collision hazard with a person walking towards the machine, that person will often simply walk away again.

2. *Issue a command to move the machine to a known safe state* This is the action taken if the machine sensors indicate that there is still a potential hazard around the machine.

Once the information from the sensor system indicates that the situation is again safe, the safety manager relinquishes command of the system and returns this to the activities manager with an indication that the operation must be restarted. Normal system operation can then continue.

Clearly, the sensor system on the machine is a safety-critical part of the system. We have not explored this in detail but have worked on the assumption that sensor malfunction is detectable by the safety manager. In the event of sensor failure or malfunction, the safety manager takes over control of the system, moves it to a safe state and suspends its operation.

# 7 Conclusions

The work that we have done has demonstrated that the notion of a safety manager which takes responsibility for ensuring system safety is viable, at least for relatively slow moving mechanical systems. We have also discovered that an understanding of the hazards that this safety management system must deal with is best tackled from a physical perspective rather than using event-based hazard analysis techniques such as fault-tree analysis and failure mode effect analysis. The principal reason for this is that we are not concerned in this analysis with system failures but with developing an understanding of hazards which can arise during the normal operation of the system. In this respect, the work is distinct from some other classes of safety-critical system where it is assumed that the principal causes of hazards are some kind of system failure.

Our original model of a safety management sub-system (i.e. the safety manager and its associated communications) was found to be simplistic. It was not realistic or indeed safe to rely on the moderation of actions by the safety manager and information from the sensors on the machine which were required for normal operations. Rather, we needed additional sensors so that the safety manager could continually check for hazards which might arise in the course of an operation. Economic factors also meant that having a dedicated communications systems for the safety manager was unrealistic so clearly the CANBUS is a safety-critical component in our system.

From the work which we have done, we have identified a number of areas which we think are worthy of further research. These include:

1. The safety implications of using CANBUS-based architectures in automotive systems.

2. Systematic techniques of hazard analysis which can be used while the system requirements are established. These cannot assume any particular design and should not be based on the assumption of system failure.

3. Further work on the concept of a safety management sub-system and an investigation into further domains where this approach may be applied.

4. Sensing systems to detect the presence of underground objects - we were not able to discover any practical means of doing this for our current version of LUCIE.

# References


Hope S. (1993). "Methodologies for Hazard Analysis and Risk Assessment in the Petroleum Industry", *Hazard Prevention*, Jul./Aug., 24 - 32

Leveson N. (1985). "Software Safety" in *Resilient Computing Systems*, Anderson, T. ed., Collins, 122 - 143

Margrave, F.W. and Seward, D.W. (1994). "Establishing the Safety Argument for the Use of Programmable Electronics Systems in Transport", 27th Int. Symp. on Automotive Technology & Automation, Aachen, Germany, 519-526

Seward, D.W., Bradley, D.A. and Margrave, F.W. (1994). "Hazard Analysis Techniques for Mobile Construction Robots", 11th Int. Symp. on Robotics in Construction, Brighton, England 35- 42

Seward, D.W., Margrave, F., Sommerville, I., and Kotonya, G. (1995). "Safe Systems for Mobile Robots - The Safe-SAM project", Achievement and Assurance of Safety, Proc of SCS Symp, Brighton, England, Springer Verlag, 153-170

Seward, D.W., Margrave, F., Sommerville, and Morrey, R. (1996). "LUCIE the Robot Excavator - Design for System Safety", Proc. IEEE Int. Conf. on Robotics and Automation, Minneapolis, U.S.A., 963-968

Seward, D.W., Quayle, S. (1996). "System Architectures and Safety for Mobile Construction Robots", 13th Int. Symp. on Robotics in Construction, Tokyo, 727-732

# Increasing Software Integrity Using Functionally Dissimilar Monitoring

David M Johnson
Department of Aerospace Engineering, University of Bristol
Bristol, UK

## Abstract

Dissimilarity or diversity has been widely used, particularly in the aerospace industry, as a means of improving the integrity of software based systems. This is primarily because of continued doubt over our ability to produce software of the integrity required for safety- or life-critical systems. However the benefit of dissimilarity has not been quantified, and it has been shown that failures of independently coded software versions may be coincident. In order to provide a better guarantee of independence between the software versions, this paper proposes the use of *Functionally Dissimilar Monitoring*, based on the analysis of system safety requirements. It also proposes that statistical methods are used to measure or estimate the reliability of each software version, and their independence. The application of these techniques requires further research, but it is hoped that the methods discussed will provide a *quantitative* basis for the certification of software in safety-critical systems.

## 1   Introduction

Software is now used extensively in safety-critical applications, despite concerns regarding our inability to accurately quantify the reliability, or integrity of the software [Bennett 92]. Work on formal methods, and on software reliability measurement and prediction continues. The pros and cons of these techniques are discussed extensively elsewhere (e.g. [Gerhart 94], [Brocklehurst 92], [Knight 94]), but neither technique yet offers a practicable and cost effective solution to the problem. Instead, software for safety-critical applications is developed according to sets of rules or guidelines (e.g. [RTCA DO-178], [Def Stan 00-55]), which essentially define good practice, but which do not attempt to quantify the reliability or integrity achieved.

In many cases software diversity or dissimilarity is employed in order to increase the integrity of the system. The concept is simple. By coding $n$ versions of the software, it is hoped that any residual errors remaining after verification and validation are complete, will not be common to all versions of the software. Discrepancies between the outputs of the different versions can be detected, or a voting scheme can be used, to prevent the erroneous outputs affecting the system. The effectiveness of this technique however relies on the assumption that the same errors will not occur in the different software versions, that is that the software versions are truly independent.

This paper examines the use of *functional dissimilarity* to provide independence between software versions. It proposes a systematic technique for applying functional dissimilarity and suggests methods that might be used to quantify the independence of the software versions. It is desirable to define a method that can be used to provide a *quantitative* measure of system integrity, of the order required for safety-critical applications. By using functional dissimilarity, and measuring the independence between the software versions, it should be possible to lower the integrity target for each version. Software reliability measurement, and related techniques, may then be used to quantify the overall integrity of the system.

## 2  Software Dissimilarity

Software dissimilarity can be achieved in a number of ways, essentially depending on the development phase in which the dissimilarity is introduced [Kelly 91]. The different methods can be combined to reduce the probability that common errors will exist in the different software versions.

At the lowest level, common source code may be executed on different processors, by compiling with different compilers (as for the Boeing 777 Flight Control Computers [Hills 96]). This allows compiler errors to be detected (unless they are common to all compilers) and provides an effective means of detecting hardware failures. Errors in the source code will however be common to all versions, and extensive verification and rigorous development methods are therefore needed in order to achieve the high integrity required for the system. For the Boeing 777, the software was developed to RTCA DO-178A guidelines for level 1 software.

At the next level, the source code may be produced independently from a common specification and design. The different software versions may be coded in different languages and/or coded by independent teams. This type of dissimilarity attempts to detect coding errors, though there is evidence that common errors may be made by independent programmers and errors in the specification or design will not be detected.

Knight & Leveson made perhaps the most important study of the assumption of independence between separately written software programs [Knight 86]. Twenty-seven separate programs were written, to the same specification, on two different sites by people of diverse background, but all with some programming experience. The programs were subjected to acceptance tests and debugged before being subjected to a more extensive set of 1,000,000 randomly generated test cases.

Program outputs and intermediate values were recorded for each test case. These results were compared with results from a 'golden' program which had been more thoroughly tested and reviewed and was taken to be correct. Discrepancies between the results from any individual program and the golden program were recorded and analysed.

The results presented in their report showed that for many test cases, more than one program failed. For 2 of the test cases, 8 programs failed and the worst correlation of test case failures between any two programs was failure of 325 out of the 1,000,000 test cases run. The worst performance from any individual program was failure of 9,656 test cases. This analysis however looks only at the number of test case failures and not at whether the actual results from the two programs were identical.

Analysis of the causes of failure, i.e. identification of the coding error, did however reveal that identical or similar errors had been made independently by different programmers, though it would appear, from analysis of the results presented, that this represented only a relatively small proportion of the correlated test case failures. Two common errors were specifically reported. One was related to a problem of precision and the false assumption, that with limited precision, comparing the cosines of two angles was equivalent to comparing the two angles directly. The second was an error in determining whether 3 points lay in a straight line, from the angle between them (the 0 degree condition was missed). This experiment has often been cited against the use of diversity as a means to improve the integrity of the system, and diversity has been criticised as "being applied with only an incomplete understanding of the proper procedures and without accurate knowledge of the benefit to be realized" [Todd 91]. Whilst this criticism may be valid, it is also clear from the experiment, and from the work of other researchers (e.g. [Bishop 86]) that some benefit, in terms of integrity, is obtained through the use of diversity. It was concluded though that "minor differences in the software development environment, such as the use of different programming languages for the different versions, would not have a major impact in reducing the incidence of faults that cause correlated failures" [Brilliant 90].

Dissimilarity may also be introduced in the design [Avizienis 84]. This may be at the lower levels of design, for example module design by independent teams, or at the higher levels. Some dissimilarity may be forced by imposing different design rules, or by providing different hardware architectures for the different software versions (e.g. a multi-processor architecture for one and a single processor for another). This reduces the probability of common design errors, but the possibility of common specification errors remains and again it is possible that the separate teams will independently make the same errors.

For complex systems, it has been found beneficial to use formalised specification languages (e.g. [Hall 96], [Briere 95], [Johnson 96a]). These may be true formal specification methods with a strict mathematical basis, or graphical representations of software requirements, so long as the specification is unambiguous. This type of specification normally defines functional requirements in greater detail than plain language specifications, because nothing can be left open for interpretation. The effect of this is that the specifications become more complex, and that they contain elements of design, or constrain the design of the software more than a traditional specification. This in turn makes the tasks of design and coding more straightforward and gives a clearer basis for verification and validation activities performed on the design and code. The end result is that more errors are made in specification, but very many less errors are made in design and coding [Johnson 96a]. In the experiment conducted by Knight & Leveson [Knight 86], the specification that the programmers used as the basis for their programs was a plain language specification. If the specification had been more formal, using a mathematical or symbolic representation of the requirements, then it is unlikely that the common errors discussed above would have been made, provided that the specification was correct. If the specification was incorrect however it is likely that all of the programmers would have made the same errors, due to the errors in the specification. Thus the application of formal specification techniques can destroy any benefit that might be claimed for the use of software diversity, if a common specification is used. Dissimilarity in design and coding therefore provides relatively little benefit, as the majority of errors, those in the specification, will remain common to the different software versions. To resolve this, we require dissimilar specifications. The problem is how to produce these dissimilar specifications.

## 2.1 Dissimilar Specification

Given that some formalised specification language is to be used, one possibility is to independently produce specifications using different formalised specification languages. This is feasible but has a number of disadvantages. It requires the use of two (or more) different specification languages. This implies increased costs in training, tool support and development. More significant though, is the problem of producing the detailed formal specifications independently, but consistently. Even if each specification satisfies the requirements, and these requirements are correct, there are likely to be small differences in behaviour due to ambiguity or imprecision in the requirements. These may be detected as discrepancies and result in failure. This is a general problem for the application of diversity, in that the greater the degree of diversity that is introduced, the more likely it becomes that disagreement between the versions will occur. A further problem is that the formal specifications will normally share a common requirements specification, and so faults in the requirement specification may propagate into the formal specifications [Kelly 91].

Another possibility is to artificially force some dissimilarity into the specifications. This may be achieved, for example, by deliberately inverting logic from one specification to another. This may help to eliminate common errors in the subsequent design and coding processes but is unlikely to eliminate common errors between the specifications as one is effectively derived directly from the other.

However, we have already seen in the discussion above that there may be a range of behaviour that satisfies the requirements of the software (and system), and that therefore there is a range of specifications that can be written, each of which will satisfy the requirements. There is also clearly system behaviour that would not meet the requirements. More specifically there are ranges of behaviour of the software that represent a hazard to the system. If we can identify these ranges, then we can define a specification based on unacceptable behaviour as a dissimilar monitor of a specification based on required behaviour. This is the principle underlying the proposals in this paper for the systematic use of functional dissimilarity.

## 2.2 Functional Dissimilarity

The principle of functional dissimilarity is that one software version can monitor the behaviour of another using an entirely different (functionally dissimilar) set of algorithms. This is best illustrated by an example.

An aircraft braking computer is required to control brake pressure, and provide anti-skid protection to prevent wheels locking and to optimise braking efficiency. The anti-skid algorithms necessary to achieve this are complex and require high iteration rates to achieve the desired performance. A functional hazard analysis for the braking system reveals two failure modes that are of particular concern, relating to the anti-skid function:

The loss of anti-skid protection, resulting in wheel locking and possible tyre burst.

The loss of braking due to inadvertent release of the brakes due to malfunction of the anti-skid control.

Monitoring of the anti-skid function could be achieved by a comparison of results from a software version running the same algorithms on another processor (or even the same processor). This however would require a lot of processing power for the monitoring task, exchange of data between the control and monitor software and close synchronisation to avoid disagreement. It would be capable of detecting certain hardware failures, and coding errors (for errors that were not common to the two versions) but would not detect specification or design errors in the anti-skid algorithms.

Functionally dissimilar monitoring algorithms can be defined based on suitable definitions of the hazards. These definitions should be as clear, precise and mathematical as possible, and should be based on analysis of the risks. For example, the loss of anti-skid protection might be defined as any wheel having a slip ratio greater than 90% (wheel speed less than 10% of the aircraft speed), for more than 50 ms, without a brake release signal greater than 30 mA being sent to the brake servovalve. Similarly, the loss of braking could be defined as a release current greater than 20 mA being sent, to half or more of the brakes, when the braking commanded is more than 50% of the maximum and the wheel slip ratios are less than 20%, for more than 100 ms (these definitions are purely illustrative - the actual definitions would be based on a quantitative assessment of the risks of tyre burst caused by deep skids, and of the loss of braking performance, and the level of hazard represented). This definition of the hazards then forms the basis of the monitoring requirements and these requirements can be defined and implemented quite independently of the basic functional requirements. The algorithms used to monitor are fundamentally dissimilar from those used to control the system and are based on a different set of requirements, those for system safety.

In order to apply this technique systematically, we need to perform a functional hazard analysis for the system, identify software failure modes that could cause or contribute to causing the hazards identified, and define monitoring requirements based on mathematical descriptions of unacceptable system behaviour. Typically this technique need only be applied for system failure modes with more severe consequences. The functional hazard analysis will identify these and the use of functional dissimilarity can be restricted to those functions with the highest criticality and therefore the highest integrity requirements. This reduces the cost of providing dissimilarity, since the dissimilar monitoring software will normally be very much smaller than the control software. The identification of these failure modes, and the associated monitoring requirements will be defined by the *System Safety Specification*.

## 2.3 System Safety Specification

The aim of the system safety specification is then to define a set of requirements on which to base a specification for functionally dissimilar monitoring. The first stage is the functional hazard analysis. This is already performed in the early stages of system design and therefore does not represent any significant overhead in the application of the proposed technique.

Requirements for system safety (e.g. [IEC SC-65A 94]) are normally expressed in probabilistic terms. That is the allowable probability of failure is defined as a function of the effect of the failure. For commercial aircraft, these allowable probabilities vary from $10^{-5}$ per flight hour for a *Major* failure (possibility of minor injuries), to $10^{-9}$ per flight hour for a *Catastrophic* failure (loss of the aircraft)[JAR 25], [FAR 25]. The functional hazard analysis examines possible failure modes of the system (or function), and their effect. It categorises failure modes according to the severity of the effect and then defines safety objectives as probabilities of occurrence of each type of failure, based on the applicable safety regulations.

This is normally followed by a preliminary safety assessment that identifies possible causes of the failure modes identified in the functional hazard analysis. This is also already performed in the early stages of system design, to verify the choice of system architecture and to allocate safety budgets or requirements to the various system components. Software failure modes can be identified using this technique, though typically the objective is only to identify the most severe effect of software failure, so that the criticality of the software (as a whole) can be determined. For the purposes of the system safety specification, software failures must be considered more carefully. Each of the hazardous failure modes identified in the functional hazard analysis must be analysed, and the possible contribution of software failure must be identified. This type of analysis has been proposed before, both as a basis for software fault tree analysis [Leveson 91] and as a basis for analysing where to concentrate effort in order to optimise the economic efficiency of testing [Sherer 91].

The next stage is to define more precisely what constitutes hazardous behaviour. For each of the hazardous software failures identified, a mathematical description of the failure must be produced. This must be based on analysis of the extent of malfunction required for the failure to reach a certain level of hazard. For example loss of 10% of braking power is less severe than loss of 50% of braking power, which is less severe than loss of all braking power, and the probability objectives for these failures would probably be different. For complex systems, in particular, it may be necessary to use system simulation to support this analysis, but again this is often an integral part of the system development process, and is in any case a valuable technique to aid system design [Johnson 96b], [Rea 93].

The System Safety Specification is then the specification for the dissimilar monitoring software. Because the specification is fundamentally different from the specification of the functional requirements, a greater degree of independence is assured [Theuretzbacher 86], but we still do not yet have a means to measure the integrity or safety of the resulting software and system. Proposals for measuring system integrity will be developed in the next section, but even if functionally dissimilar monitoring is not used, development of a System Safety Specification may prove valuable and can be used to complement reliability measurement and estimation techniques and proof methods. Firstly, having identified the possible hazardous failure modes, we can analyse the errors detected during development of the control software, and identify those errors that could result in hazardous system behaviour. It may then be possible to perform a useful statistical analysis of the proportion of failures that are hazardous. Since experience shows that the majority of software failures are not hazardous, this analysis may provide an estimate of software safety, greater than the estimate of software reliability. Another possibility is to attempt to prove that the specification of the control software and/or the software itself cannot exhibit the hazardous behaviour defined in the System Safety Specification, for example using software fault trees [Leveson 91]. There is usually only a relatively small number of feared events to be analysed, and thus by verifying safety separately we can reduce the size of the problem and hence simplify the required proof [Leveson 84].

# 3  Measuring Integrity

The use of a system safety specification, and the systematic use of functional dissimilarity to detect the more severe failure modes of the software, can be expected to improve integrity, but the objective set out at the start of this paper was to be able to *quantify* the integrity. In order to do this, we need some measure of the reliability of the control software and of the ability of the functionally dissimilar monitoring software to detect hazardous behaviour. Alternatively, we need measures of the reliability of each software and of their independence.

Software reliability measurement uses statistical testing techniques to quantify reliability. The software is tested using random test cases, usually drawn from a test distribution weighted according to the operational profile of the software. If the software passes all of a number of tests, then a certain level of reliability can be stated with a given level of confidence. The principal problems with this technique (and therefore the main areas being researched) are the large number of tests required to achieve a high degree of confidence that the software has a high level of reliability, the definition of test cases and the need to have an automatic check of whether each test has passed or failed, i.e. the need for an 'oracle' [Lyn 92], [May 95].

Simple reliability *measurement* for safety-critical software, where the failure objective may be as low as one failure in $10^9$ hours, is not feasible. However if the critical system failure requires two independent software failures then the objective can be reduced and reliability measurement techniques become a little more attractive. To gain 99% confidence that the probability of failure of the software was better than $10^{-9}$ would require the software to successfully pass in excess of 4.6 billion test cases (using simple probability theory). This, though simplistic, demonstrates that such an approach is not viable, and [Butler 93] have shown, more comprehensively, that the quantification of life-critical software by experimental means is infeasible. To establish the same level of confidence that each of two pieces of software had a probability of failure lower than $10^{-5}$ would require around 460000 tests each, giving 920000 tests in total, a reduction by a factor of 5000. However we also have to establish the independence between the software versions and to define methods that will allow us to automate the generation of test cases and the checking of results, as the quantity of testing required will remain high.

There is also the problem, that in principle, if the software fails any of these tests, then the cause of the failure should be corrected and the entire set of tests should be repeated. Often this is not done, as it is argued that the changes made to one part of the software will not affect other parts. A limited set of tests is conducted to show that the error has been corrected, and that the software in the region of the modification has not regressed. Alternatively, if the failure is considered minor (not affecting safety, or significantly affecting system operation), it may be accepted and not corrected at all.

Various other reliability estimation methods have been researched in an attempt to overcome some of these problems. These include *reliability prediction* or *reliability growth* techniques, which collect test failure data during development of the software, and extrapolate failure rates to provide an estimate of future reliability [Littlewood 90]. Methods of reliability estimation based on *accelerated testing*, *complexity analysis* and on *test coverage analysis*, have also been proposed. Methods of measuring software *testability* have also been put forward, which may aid or augment the results of reliability analysis, or be useful in guiding software design and testing [Voas 95]. Bertolino provides a useful overview of these techniques, and discusses some of their problems and limitations [Bertolino 95]. Researchers have also looked at combining some of these techniques, for example [May 95], which effectively combines reliability measurement and coverage analysis (or partitioning) techniques and [Hamlet 94] which combines reliability measurement based on random testing, with testability analysis. These methods have the potential to produce higher reliability estimates from a reduced quantity of testing, but require further research before they can be applied with confidence, as a

224

quantitative basis for the certification of safety-critical software. In any case, a safety target of $10^{-9}$, or similar, is likely to remain difficult to reach, and so the use of functional dissimilarity remains attractive.


## 3.1 Testing the Software

For the control software, test cases may be selected randomly according to the operational profile of the software, in order to measure the reliability. This is not straight forward, but is the subject of on-going research [Musa 93], and some debate as to the validity of assuming any particular operational profile [Hamlet 96]. For the monitor software, the problems are different however. Generating test cases according to the operational profile will produce large numbers of test cases representing correct operation of the control software, and very few test cases representing erroneous behaviour. This is not very useful since the tests will tell us little about the ability of the monitor software to detect failures of the control. A further problem is that we have no means to identify the distribution of erroneous outputs from the control software, in order to provide a full operational profile for the monitor software.

The problem of selecting tests for the monitor software in this case, is similar to the problem of defining tests for software that's sole purpose is monitoring. Software for monitoring nuclear power plants, or industrial processes falls into this category, and in the same way, it is not useful to expend thousands of hours of test effort checking that nothing happens when nothing should happen (though some testing of this type is necessary). The tests for monitoring software of this type can however be based on known failure modes and known (or estimated) failure rates for the components of the system being monitored. An estimate of the *probability of failure on demand* (pfd) can then be obtained. The process of defining test cases is not simple, and may require the use of elaborate simulations of the monitored system (environment simulation), but suitable techniques have been defined and developed [May 95]. Environment simulation may also play a role in generating test cases for the control software, since more realistic input profiles may be generated if the software is placed within a simulated environment and test cases are generated using random inputs to the environment. The problem of software monitoring software is different however, as little is known about the possible failure modes of the control software or their probability of occurrence.

One possibility to tackle this problem is to develop the monitor software first, to test it using conventional test methods to try to eliminate errors, and then to use it to monitor the behaviour of the control software during its testing. The control software will at this stage of its development contain errors and these, if they result in hazardous behaviour, should be detected by the monitor software. The proportion of such failures detected can be used to provide an estimate of the probability that the monitor software will detect any such failure. This is unlikely to provide sufficient data however, and failure of the monitor software to detect any of the hazardous failures will further reduce the reliability estimate that can be made.

Another possibility is to deliberately introduce errors in the control software and to check the ability of the monitor software to detect the hazardous behaviour that may result. Fault injection (or mutation) techniques have been used before in reliability and test coverage analysis (e.g. [Thévenod-Fosse 94], [Fleyshgakker 94]), but this approach is also flawed, for this application, since errors introduced deliberately may be unrepresentative of the real errors in the software, and they will not necessarily result in hazardous behaviour. Mutations are normally made by making single, syntactically correct changes to the source code, but this does not cover all possible types of error, and introducing errors that may cause hazardous behaviour will require a thorough understanding of the operation of the software. There may also be difficulty in defining test cases, even for those errors that can cause hazardous failure, that will stimulate them into exhibiting such behaviour and the whole process would be extremely difficult to automate.

Since nothing is known about the residual errors in the control software, the failure modes that may result or their relative failure rates, testing the monitor software using a random selection of hazardous behaviour is as fair a method as any. If we define the input state of the control software as being the combination of inputs and internal state at any instant, then there is a region of the possible output space that represents correct operation (compliant with the requirements), and a region of the output space that represents incorrect operation. Within this second region, there is a region that represents hazardous behaviour. We can therefore choose inputs (and hence internal states which are computed from the inputs) according to the operational profile, and outputs randomly from the hazardous output space, and apply these as test cases for the monitor software. It has been argued that the use of a uniform random test distribution, in any case, has more validity than the assumption of any particular operational profile [Hamlet 96], and so consideration may also be given to using a uniform random test distribution for the control software. It is easy to automate the generation of random test cases, but for the monitor software we only wish to select test cases that should be detected as hazardous states by the monitor. We will therefore need a method of deriving the hazardous output space, from the system safety specification, so that the generation of monitor test cases can also be automated. It will not be viable (or random) to generate test cases manually. The proposed method of measuring the integrity of the software, to provide a quantitative basis for certification is described below.

## 3.2 A Quantitative Basis for Certification

First, we subject an immature version of the control software to random testing and we detect $F_c$ failures in $N$ tests. We can analyse these failures in order to establish their cause (the error) and to assess their criticality. Those failures that represent a hazard to the system we term *Hazardous Failures* and we detect $f_c$ such failures. We can use this data to estimate the probability $\lambda_h$ that when the software fails, it fails in a hazardous manner. An upper bound on this probability can be given with a statistical confidence $\mu_h$ according to:

$$(1-\mu_h) \le \sum_{i=0}^{f_c} \lambda_h{}^i (1-\lambda_h)^{F_c - i} \binom{F_c}{i}$$

We can also submit an immature version of the functionally dissimilar monitor software to random testing. We can use the same random input profile as before and generate control outputs randomly from the hazardous output space. In addition we should test the monitor software, with the control software, for those input profiles that resulted in hazardous failure of the control software. The monitor software should detect the failure in most cases. If, for the random tests, it fails to detect $f_m$ failures (i.e. the monitor software fails $f_m$ times) out of $N$ tests and for the tests with the control software it fails to detect $f_{mc}$ failures of the $f_c$ hazardous control software failures, then we can make a statistical estimation of the likelihood of independence or of a level of confidence $\mu_k$ in a dependence factor $k$. The dependence factor $k$ is the factor by which the overall probability of failure on demand (pfd), of the monitor software, $\lambda_m$, is multiplied, to give an estimate of the pfd of the monitor software, for actual hazardous failures of the control software. In order to ensure that this analysis is pessimistic (safe) we must calculate the minimum value of $\lambda_m$ that could reasonably result in $f_m$ failures, and the maximum value of $k\lambda_m$ and hence $k$ that could result in only $f_{mc}$ failures. Again, the values calculated will depend on the confidence levels required. The product of the confidence levels will give the overall confidence $\mu_k$ in the value of $k$.

If we then assume that these properties $\lambda_h$ and $k$ remain constant as the software gains maturity and errors are corrected, then when we submit the mature software versions to random testing (which should reveal no failure) then we can calculate an overall integrity figure for the software, based on their individual reliability estimates and these two factors. If we have confidence $\mu_c$ that the control software failure probability is less than $\lambda_c$ and confidence $\mu_m$ that the monitor software failure probability is less than $\lambda_m$, then we can estimate the overall integrity as:

$$\lambda_c \cdot \lambda_m \cdot \lambda_h \cdot k \text{ with a confidence of } \mu_c \cdot \mu_m \cdot \mu_k \cdot \mu_h$$

The following numerical example illustrates the application of the method.

Suppose the random testing of the immature control software reveals 504 failures, 17 of which are assessed as hazardous. If we aim for 99% confidence in the estimate of $\lambda_h$ then:

$$0.01 \le \sum_{i=0}^{17} \lambda_h{}^i (1-\lambda_h)^{504 - i} \binom{504}{i}$$

This equation is not easily solved, but may be solved numerically with the aid of a simple program, to give a value of $\lambda_h$ of 0.05745.

Now if the random testing of the monitor software reveals 243 failures in 20000 random tests and detects 15 of the 17 control failures, then we calculate $k$ and $\mu_k$ as follows. We will aim for 99% confidence at each stage, and we must first calculate the *minimum* value of $\lambda_m$ from:

$$0.99 < \sum_{i=0}^{243} \lambda_m{}^i (1 - \lambda_m)^{20000-i} \binom{20000}{i}$$

Solving this equation gives 99% confidence that the value of $\lambda_m$ is no lower than 0.01046.

We then calculate the *maximum* value of $k\lambda_m$ from:

$$0.01 \leq \sum_{i=0}^{2} (k\lambda_m)^i (1 - k\lambda_m)^{17-i} \binom{17}{i}$$

This gives the maximum value of $k\lambda_m$ as 0.40993 and hence the maximum value of $k$ as 39.2, with approximately 98% confidence.

If we then, through conventional statistical estimation of the reliability of the mature control and monitor software (or by some other means) estimate failure rates for each of $10^{-5}$, with a confidence of 99%, then the overall integrity estimate, i.e. the probability of a hazardous event occurring, is less than:

$10^{-5}$ x $10^{-5}$ x 0.05745 x 39.2 with a confidence of 0.99 x 0.99 x 0.98 x 0.99

i.e. $< 2.252$ x $10^{-10}$ with a confidence of 95%

Thus we have used statistical measures of the proportion of failures resulting in hazardous behaviour and of the independence of the control and monitor software in order to enhance our integrity estimate for the system. The method may also be adapted for analysis of conventional n-version software integrity. It provides the prospect of a quantitative basis for certification, against very high integrity requirements, but so far as it is statistically based, and relies on the validity of the assumptions made, it does not provide an absolute guarantee.


# 4  Discussion

The assumptions made in the above analysis are that the statistical measurement of software reliability is a valid measure of real reliability, and that the proportion of control software failures that are hazardous, and the degree of dependence between control and monitor failures remain constant, as errors are removed from the software. There is also an implicit assumption, that analysing the integrity of the system in this way will produce the desired result.

Since development effort often concentrates (deliberately) on the detection and removal of errors from critical areas of the software [Sherer 91], it seems reasonable to expect that the proportion of hazardous failures will not increase. Similarly, although the use of particular testing methods may fail to uncover particular types of fault, there is no reason to suppose that a higher proportion of the residual faults will result in common failure of the control and monitor software. We would expect, given that they are developed independently from entirely separate requirements, that they would be truly independent ($k = 1$). The method of assessing dependence given above will give a value of $k$ greater than 1, even when the proportion of actual faults detected by the monitor is equal to the overall proportion of faults detected. It gives therefore, in some sense, a pessimistic estimate of the dependence between the control and monitor. Nevertheless, the validity of these assumptions requires further investigation.

The use of statistical reliability measures, and other methods of assessing reliability are the subject of much research, but even if some doubts remain over their use, we can use the methods described above to provide a quantitative basis for certification, in addition to the current good-practice guidelines. The possibility of failure to demonstrate the required integrity levels, using the methods described, must however be addressed. If the dependence factor $k$ is very high, and/or the proportion of failures that are hazardous is high, this will increase the number of random tests of the mature software versions necessary to show a particular level of integrity. It will be necessary to investigate ways of preventing the required quantity of testing from becoming excessive. This, and the investigation of the assumptions discussed above, will be the subject of further research.


# 5 Conclusion

This paper has identified a systematic methodology for defining functionally dissimilar monitoring software, to protect a system from failure of the control software (assuming that a fail-safe state exists). This monitoring software is based on the analysis of system hazards that may be caused by the control software. A system safety specification is derived from analysis of system safety requirements and this forms the basis for the monitor software requirements. Since the control and monitor software will be developed independently, from dissimilar specifications, it is expected that a very high degree of independence will be achieved. It is desirable however, for safety-critical systems, to be able to quantify the integrity of the system software. A method based on reliability measurement has been proposed. This uses measurements of failure correlation between the control and monitor software and of the proportion of control software failures that are hazardous, in immature versions of the software, together with reliability measurements of the mature software versions. Although this will not provide a guarantee of integrity, and will not replace the need for good development practices, it is hoped that it will provide a quantitative basis for the future certification of high-integrity or safety-critical systems.

# References


[Avizienis 84]      Avizienis, A. & Kelly, J.P.J. Fault Tolerance by Design Diversity: Concepts and Experiments. IEEE Computer August 1984 (IEEE) pp 67-80.

[Bennett 92]      Bennett, P.A. Evolving Standards for Safety Critical Software. Proceedings of Aerotech '92 C428/24/142 (IMechE) 1992.

[Bertolino 95]      Bertolino, A. Software Testing for Dependability Assessment. Objective Software Quality: Proceedings of the Second Symposium on Software Quality and Acquisition Criteria May 1995 (Springer) pp 236-248.

[Bishop 86]      Bishop, P.G., Esp, D.G., Barnes, M., Humphreys, P., Dahll, G. & Lahti, J. PODS - A Project on Diverse Software. IEEE Transactions on Software Engineering Vol. 12, No. 9 September 1986 (IEEE) pp 929-940.

[Briere 95]      Briere, D., Ribot, D., Pilaud, D. & Camus, J-L. Method and Specification Tools for Airbus Onboard Systems. Microprocessors & Microsystems Vol. 19, No. 9 September 1995 pp 511-515.

[Brilliant 90]      Brilliant, S.S., Knight, J.C. & Leveson, N.G. Analysis of Faults in an N-Version Software Experiment. IEEE Transactions on Software Engineering Vol. 16 No. 2, February 1990 (IEEE) pp 238-247.

[Brocklehurst 92]      Brocklehurst, S. & Littlewood, B. New Ways to Get Accurate Reliability Measures. IEEE Software July 1992 (IEEE) pp 34-42.

[Butler 93]      Butler, R.W. & Finelli, G.B. The Infeasibility of Experimental Quantification of Life-Critical Software Reliability. IEEE Transactions on Software Engineering Vol. 19, No. 1 January 1993 (IEEE) pp 3-12.

[Def Stan 00-55]      Defence Standard 00-55 (MoD) The Procurement of Safety Critical Software in Defence Equipment 1991.

[FAR 25]      FAR 25.1309 Federal Aviation Regulations (+ Advisory Circular AC25.1309-1A, "System Design and Analysis", 1988).

[Fleyshgakker 94]      Fleyshgakker, V. & Weiss, S. Efficient Mutation Analysis: A New Approach. Proceedings of the 1994 International Symposium on Software Testing and Analysis, August 1994. Software Engineering Notes Special Issue (ACM) pp 185-195.

[Gerhart 94]      Gerhart, S., Craigen, D. & Ralston, T. Experience with Formal Methods in Critical Systems. IEEE Software January 1994 (IEEE) pp 21-28.

[Hall. 96]      Hall, A. Using Formal Methods to Develop an ATC Information System. IEEE Software March 1996 (IEEE) pp 66-76.

[Hamlet 94]      Hamlet, D. Foundations of Software Testing: Dependability Theory. Proceedings of the Second ACM SIGSOFT Symposium on Foundations of Software Engineering. Software Engineering Notes December 1994 (ACM) pp 128-139.

[Hamlet 96]      Hamlet, D. Predicting dependability by Testing. Proceedings of the 1996 International Symposium on Software testing and Analysis (ISSTA) Software Engineering Notes May 1996 (ACM) pp 84-91.

[Hills 96]      Hills, A.D. The Primary Flight Computer for the Boeing 777 - a Description. GEC Review Vol. 11, No. 1 1996 (GEC) pp 11-20.

[IEC SC65A 94]      IEC SC65A Functional Safety - Safety Related Systems, Parts 1-3, (committee draft) 1994.

230

[JAR 25]  JAR 25.1309  Joint Airworthiness Requirements Change 14 May 1994 (+ Advisory Material AMJ 25.1309, "System Design and Analysis, Advisory Material Joint").

[Johnson 96a]  Johnson, D.M.  The Systems Engineer and the Software Crisis. Software Engineering Notes March 1996 (ACM) pp 64-73.

[Johnson 96b]  Johnson, D.M.  A Review of Fault Management Techniques Used in Safety-Critical Avionic Systems.  Progress in Aerospace Sciences (Elsevier Science Ltd.) - To Be Published.

[Kelly 91]  Kelly, J.P.J., McVittie, T.I. & Yamamoto, W.I.  Implementing Design Diversity to Achieve Fault Tolerance.  IEEE Software July 1991 (IEEE) pp 61-71.

[Knight 94]  Knight, J. & Littlewood, B.  Critical Task of Writing Dependable Software. IEEE Software January 1994 (IEEE) pp 16-20.

[Knight 86]  Knight, J.C. & Leveson, N.G.  An Experimental Evaluation of Failure Independence in Multi-Version Programming.  IEEE Transactions on Software Engineering Vol. 12 No. 1, January 1986 (IEEE) pp 96-109.

[Leveson 84]  Leveson, N.G.  Software in Computer-Controlled Systems.  IEEE Computer February 1984 (IEEE) pp 48-55.

[Leveson 91]  Leveson, N.G., Cha, S.S. & Shimeall, T.J.  Safety Verification of ADA Programs using Software Fault Trees.  IEEE Software July 1991 (IEEE) pp 48-59.

[Littlewood 90]  Littlewood, B.  Modelling Growth in Software Reliability.  Software Reliability Handbook  Ed. Rook, P. (Elsevier Science).

[Lyn 92]  Lyn, M.R. & Nikora, A.  Applying Reliability Models More Effectively. IEEE Software July 1992 (IEEE) pp 43-52.

[May 95]  May, J., Hughes, G. & Lunn, A.D.  Reliability Estimation from Appropriate Testing of Plant Protection Software.  Software Engineering Journal Vol 10, No. 6 November 1995 (IEE/BCS) pp 206-218.

[Musa 93]  Musa, J.D.  Operational profiles in Software Reliability Engineering. IEEE Software March 1993 (IEEE) pp 14-32.

[Rea 93]  Rea, J.  Boeing 777 High Lift Control System.  IEEE Aerospace and Electronic Systems August 1993 (IEEE) pp 15-21.

[RTCA DO-178]  RTCA DO-178 Software Considerations in Systems and Equipment Certification (DO-178A 1985, DO-178B.5 Nov 1991).

[Sherer 91]  Sherer, S.A.  A Cost-Effective Approach to Testing.  IEEE Software March 1991 (IEEE) pp 34-40.

[Todd 91]  Todd, J.R. & Yount, L.J.  Digital Flight Control Systems: Some New Commercial Twists.  Proceedings of the 10th Digital Avionics Systems Conference October 1991 (IEEE/AIAA) pp 79-84.

[Theuretzbacher 86]  Theuretzbacher, N.  Using AI Methods to Improve Software Safety. SAFECOMP '86 Proceedings of the Fifth IFAC Workshop October 1986 (Pergamon Press) pp 99-105.

[Thévenod-Fosse 94]  Thévenod-Fosse, P., Mazuet, C. and Crouzet, Y.  On Statistical Structural Testing of Synchronous Data Flow Programs.  Proceedings of the First European Dependable Computing Conference October 1994 Eds. Echtle, K., Hammer, D. & Powell, D. (Springer-Verlag LNCS No. 852) pp 250-267.

[Voas 95]  Voas, J.M. & Miller, K.W.  Software Testability: The New Verification. IEEE Software May 1995 (IEEE) pp 17-28.

# Timing Aspects of Fault Tree Analysis
# of Safety Critical Systems


Janusz Górski and Andrzej Wardziński
Franco-Polish School of New Information and  Communication Technologies (EFP)
Poznań, Poland


## Abstract


In the paper we present a new approach to the analysis of Fault Trees with a particular stress on the timing aspects. The proposed method assumes that a (conventional) Fault Tree is first formalised and then is subjected to various analyses which aim at discovering if, with the given time dependencies among events, the hazard is still reachable. The method is illustrated with respect to a common case study - a gas burner system.


## 1. Introduction

Conventional  Fault Tree Analysis (FTA) concentrates on discovering possible scenarios through which hazards can develop in the system as a chain of cause-effect dependencies among events. The most common analysis performed with respect to a fault tree is based on calculations of the top event probability,  given the probabilities of the basic events in the tree.  One of the drawbacks of such analysis is that it does not cover timing relations between events which sometimes can play important role while deciding if a given combination of events can have a dangerous effect (e.g. the effect of opening the gas valve and generating a spark in a gas burner system clearly depends on timing of the two events, and can vary from 'no effect' to 'explosion').

In the paper we present an extension of the conventional FTA which provides for including time to the analysis process.  The fault tree semantics is introduced which represents the meaning of events of the tree in terms of time. Then two analytical approaches are proposed. In the first approach a fault tree is analysed top-down and the goal is to identify the *enabling condition* which represents timing dependencies between the events belonging to a given  minimal cut set. Then, if we can guarantee that this condition is continuously maintained *False* then we exclude the hazard occurrence as an effect of this particular cut set. This provides for identification of additional requirements for software, if some of the events of the cut set are under software control. In the second  approach, the tree is analysed bottom-up to find out if, given the time relations between the events in the tree, the top event (hazard) is possible in the logical sense. We use the Time Petri Net representation of the tree to support this kind of analysis. The use of the proposed techniques is illustrated in the paper by referring to a common case study (a gas burner system).

# 2. The method of analysis

The method of Time-based Fault Tree Analysis (TFTA) which is presented in this paper comprises the following steps:

**STEP 1:  Fault Tree Analysis (conventional)**

This is the traditional system-level Fault Tree Analysis conducted by an expert in safety and in the application domain. The result of this step is a set of Fault Trees, each representing a possible hazardous event together with its scenario expressed in terms of the contributing events and their causal relationships.

**STEP 2:  Fault Tree Formalisation**

The objective of this step is to define a  Fault Tree in a formal way. This removes ambiguities and provides for formal analysis of the tree.  The formalisation is based on the formal model given in [BCG90, Górski94]. The model provides for specification of  timing of the events in the tree. The formalisation  is a systematic task, consisting of a few steps as described in [GW95].

**STEP 3:  Minimal Cut Sets with Time Analysis**

This step consists of the minimal cut set analysis.  For each cut set, the time relations between the events belonging to the cut set are derived from the tree in the form of a formal expression. This time relation is expressed in the form of *enabling condition* for the cut set. If the enabling condition of each cut set of the tree is *False*, we can conclude that no cut set can contribute to the hazard occurrence.

In case the enabling condition can become *True*, we should look for possible design changes which would guarantee that the condition is maintained *False*. In particular, if the enabling condition  refers to the  *software dependent events* then the hazard can be prevented if, throughout the system operation, software maintains the negation of the whole condition. This observation provides for derivation of additional  safety requirement to be fulfilled by software [GW96].

**STEP 4:  Time Petri Net Analysis**

The objective of this step is to perform time analysis of hazard reachability. The Fault Tree is transformed into Time Petri Net and then is analysed [GMW95] with respect to the reachability of the top event. This type of analysis serves two purposes:

- is capable to discover some anomalies in time dependencies resulting from possible interference of events from different cut sets. In this case it strengthens the analysis performed in STEP 3.

- is capable to cross-check the results of the analysis of STEP 3 and by this it increases the level of assurance of the whole approach.

In the sequel, we present the subsequent steps of our approach in more detail. The presentation is given referring to a simple case study - the gas burner system.

# 3. Gas burner system

The example considered in this paper is a gas burner system presented on Fig. 1. Its mission is to supply heat to a technical process. The main component of the system is a burning chamber together with the associated gas supplying pipe and the ignition device. The valve on the gas supplying pipe and the ignition device are controlled by a computer. The computer receives external signals (e.g. from the system operator) to start and to stop the heating process.

Fig. 1. The gas burner problem.

During further analysis it is assumed that the increase of the gas concentration in the burning chamber is proportional to the time period during which the valve remains open (assuming that there is no ignition). If the valve closes, the gas concentration drops to zero immediately, due to ventilation. It is also assumed that the minimal concentration sufficient to gas explosion is known. The ignition device generates sparks in response to the command received from the controller. It is assumed that during normal operation of the system, gas is always available. The above assumptions are made to keep the case study simple enough to be tracable within the limits of this paper.

# 4. Conventional Fault Tree Analysis

A fault tree related to the gas burner example is shown in Fig.2. Fault Tree Analysis is a top-down method which starts with a *hazard* and then identifies its possible scenarios. In the gas burner system, the hazard is an event of dangerous concentration of gas (which may lead to explosion). Each event is represented as a block with the symbolic name and the textual explanation inside. The hazard of the system is represented by the top-event. In the tree in Fig.2. the top event is caused by the events E1 and E2. Both E1 and E2 have to occur in order to cause the top event, which is represented by the AND gate, G1, in the tree. The gate symbols used in the Fault Tree are similar to the symbols used in electronics. G1 is an example of the gate of *AND* type: it requires the common occurrence of both input events to cause the output event. G3 is of the *OR* type: the occurrence of any of the input events suffices to cause the output: the gas valve is open (E3) if OPEN-VALVE command was issued (E4) *OR* the valve fails (E5).

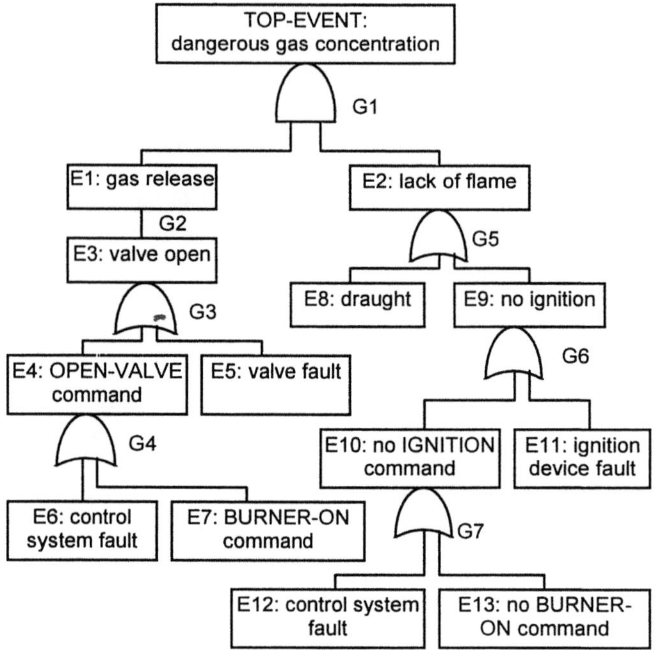



Fig. 2. A Fault Tree of the gas burner.

The Fault Tree is presented in a graphical form and the semantics of events are defined in a natural language. This leads to difficulties in defining time relations between events in the tree, e.g. it is not distinguished that the common occurrence of *gas release* and *lack of flame* can be a perfectly normal situation or a dangerous one depending on the timing of the involved events.

## 5. Fault Tree Formalisation

We use the CSDM (Common Safety Description Model) model to define the Fault Tree in a precise way. For the full definition of CSDM one should refer to [Górski94]. The systematic process of converting a conventional Fault Tree into the formally defined one is described in [GW95].

The basic semantic notion of CSDM is *event*. An event is a distinguished state of a system which can last for some time (an event has a non-zero duration). An event can have many occurrences (the same state can be repeated many times). To distinguish among them we use the notion of *action*. Action is an instance of the event. We define the $\phi$ function which gives all possible actions of an event. Action $e$ is an instance of event $E$ iff $e \in \phi(E)$. The starting and ending points of an action are called *transitions*.

To model system behaviours we introduce *Time* function. It maps transitions into real numbers denoting moments of time when those transitions occur. Each *Time* function represents one scenario of system behaviour. To check if action $e$ occurs within scenario *Time,* the $occur_{Time}(e)$ predicate can be used. For each action $e$, *start(e)* and *end(e)* represent the moments in *Time* associated with the starting and ending transitions of $e$. We can calculate duration of action $e$:

*duration(e) = end(e) - start(e)*

Two actions $a$ and $b$ can overlap in *Time:*

$overlap_{Time}( a, b ) \Leftrightarrow \exists\, t : \mathbf{Cod}(Time) \cdot start(a) < t < end(a) \wedge start(b) < t < end(b)$

During the fault tree formalisation process we refer to the system structure. Each event is defined in terms of the state of the system components. The notation used is similar to VDM [Jones90]. The components of the gas burner are listed in the table below:

| Component | Possible states | Comment |
|---|---|---|
| BURNER | IDLE<br>FIRE | the state of the burning<br>chamber |
| VALVE | OPEN<br>CLOSED | the gas valve position |
| GAS | PRESENT<br>ABSENT | the presence of gas<br>in the burning chamber |
| IGNITION DEVICE | SPARK<br>IDLE | state of the ignition device |
| IGNITE SIGNAL | ON<br>OFF | ignition signal from<br>the control system |
| VALVE SIGNAL | ON<br>OFF | valve control signal<br>(valve is to be open when<br>the signal is ON) |
| SYSTEM CONTROL<br>SIGNAL | ON<br>OFF | signal from the system<br>operator controlling the<br>heating process |

The following constants are used in the gas burner specification:

$t_G$ - time necessary for reaching dangerous gas concentration (in the absence of a spark)

$t_Z$ - time needed to open the valve (due to inertia of the mechanical parts)

The formal model of the gas burner system is as follows:

values $t_G$, $t_Z$

```
system GAS-BURNER of
    BURNER              : { IDLE, FIRE }
    VALVE               : { OPEN, CLOSE }
    GAS                 : { PRESENT, ABSENT }
    IGNITION_DEVICE     : { SPARK, IDLE }
    IGNITE_SIGNAL       : { ON, OFF }
    VALVE_SIGNAL        : { ON, OFF }
    CONTROL_SIGNAL      : { ON, OFF }
```

The events of the Fault Tree are specified in terms of the above model. As an example let us consider the following definition of the TOP-EVENT:

$$\text{TOP-EVENT}(\textit{Time}, t) \equiv \exists\, a \in \phi(\text{PTE}) \cdot \textit{duration}(a) > t_G \wedge \textit{start}(a) < t < \textit{end}(a)$$

where PTE is defined as follows:

$$\text{PTE}(\textit{Time}, t) \equiv (\,(\text{GAS} = \text{PRESENT}) \wedge (\text{BURNER} = \text{IDLE})\,)/(\textit{Time}, t)$$

The above definition introduces an auxiliary event PTE which occurs at any time moment when GAS is PRESENT (there is a gas release) and BURNER is IDLE (there is no flame - the gas is not burning). TOP-EVENT is then defined as coinciding with those occurrences of PTE which have their duration longer then $t_G$ ($t_G$ denotes the time necessary for reaching the dangerous gas concentration). Informally, the definition reads as follows: for any behaviour *Time*, TOP-EVENT holds at a time moment $t$ iff there exists an action $a$ of the event PTE such that duration of $a$ is greater than $t_G$ and $a$ holds at $t$.

Fault tree gates define causal and time relations between events. The graphical representation is informal and leaves room for many ambiguities. For example let us observe that the events E1: *gas release* and E2: *lack of flame* have to be related in time to cause a hazard. Short period of gas release in the presence of the lack of flame is perfectly normal and always occurs, before the ignition takes place. The formal approach proposed in [Górski94] allows to specify causal and time relations in a precise and unambiguous way. For instance, the meaning of gate G1 of Fig.2. is given as follows:

$$
\begin{aligned}
&\mathbf{M}(\,\text{G1}(\,\text{E1}, \text{E2}, \text{TOP-EVENT}\,)\,) \equiv \\
&\quad \forall\, \text{te} \in \phi(\text{TOP-EVENT}) \cdot \textit{occur}(\text{te}) \Rightarrow \\
&\qquad \exists\, \text{e1} \in \phi(\text{E1}),\, \text{e2} \in \phi(\text{E2}) \cdot \\
&\qquad\qquad \textit{occur}(\text{e1}) \wedge \textit{occur}(\text{e2}) \wedge \\
&\qquad\qquad \textit{overlap}(\,\text{e1}, \text{e2}\,) \wedge \\
&\qquad\qquad \textit{duration}(\text{e1}, \text{e2}) > t_G
\end{aligned}
$$

The definition says that in order to cause TOP-EVENT both E1 and E2 have to occur and overlap for a time period not shorter than $t_G$.

The result of the above process is a formal definition of the whole Fault Tree describing in a precise and unambiguous way the participating events and their relationships.

## 6. Minimal Cut Set Analysis

One of the analyses widely applied to Fault Trees in an industrial practice is the *minimal cut set analysis*. A minimal cut set is the smallest possible set of events which, if occur together, can cause a hazard. The hazard will not happen if at least one of the events of the minimal cut set does not occur. The method consists of a simple step of identification of the sufficient causes of the output event of a single gate. This step is applied repeatedly for each gate down in the Fault Tree, starting from the top gate. To illustrate this let us refer to our gas burner example. The top gate in the gas burner Fault Tree is of AND type, and therefore the top minimal cut set is:

> { E1, E2 }.

The next gate (G2) asserts that to cause E1 we need E3, and the cut set becomes:

> { E3, E2 }.

G3 is of OR type (E3 is caused by E4 OR E5), and therefore our cut set splits into two:

> { E4, E2 } and { E5, E2 }

The process is repeated for all other gates down to the leaves of the tree. The resulting minimal cut sets for the tree of Fig.2. are given below:

$\{E5, E8\}, \{E5, E11\}, \{E5, E12\}, \{E5, E13\}$
$\{E6, E8\}, \{E6, E11\}, \{E6, E12\}, \{E6, E13\}$
$\{E7, E8\}, \{E7, E11\}, \{E7, E12\}, \{E7, E13\}.$

From the above we know that if e.g. E7 and E12 occur together, they can be the cause of the hazard. However this kind of analysis leaves the interpretation of 'together' unspecified. In particular we do not know if the participating events have to occur at the same time or in some specified order, and what their durations should be. To deduce that we extend the method of minimal cut sets with additional time analysis. We assume that the common occurrence of the events E1 and E2 of a minimal cut set of two events can be formally characterised as follows:

$$\exists\, e1 \in \phi(E1),\; e2 \in \phi(E2)\,\cdot$$
$$occur(e1) \wedge occur(e2) \wedge$$
$$enabling\text{-}condition(e1, e2)$$

To illustrate the assumption let us consider the gate G1. It provides us with a minimal cut set {E1, E2} and the enabling condition which can be derived directly from the (formal) definition of the gate is as follows:

$$enabling\text{-}condition(e1,e2) =$$
$$occur(e1) \wedge occur(e2) \wedge overlap(e1, e2) \wedge duration(e1, e2) > t_G \qquad (*)$$

From the above condition it results that in order to cause the hazard, E1 and E2 should occur together, and they should overlap for at least the time period of the length $t_G$. By replacing the events E1 and E2 by their causes we can go down the tree in order to identify the condition associated with each subsequent cut set, and finally we end with the cut set composed of the leaves of the tree. For example, for the cut set {E7, E12} the associated enabling condition is as follows:

$$occur(e7) \wedge occur(e12) \wedge overlap(e7, e12) \wedge duration(e7, e12) > t_G + t_Z$$

The above condition has to be fulfilled if E7 and E12 are to contribute to the system hazard. Referring to the formal definition of the participating events we can express the above condition in the following way:

238

$\exists\ ov \in \phi(\text{OPEN-VALVE-SIGNAL-EV})$,
  $ni \in \phi(\text{NO-IGNITE-SIGNAL-EV})$,
  $cs \in \phi(\text{CONTROL-SIGNAL-EV})$ ·
  $start(ni) < start(cs) \land start(cs) \le start(ov) \land$
  $min(\ end(ov),\ end(ni),\ end(cs)\ ) - max(\ start(ov),\ start(ni),\ start(cs)\ ) \ge t_G + t_Z$

where

OPEN-VALVE-SIGNAL-EV( $Time, t$ ) $\equiv$ ( VALVE_SIGNAL = ON )/( $Time, t$ ),
NO-IGNITE-SIGNAL-EV( $Time, t$ ) $\equiv$ ( IGNITE_SIGNAL = OFF )/( $Time, t$ ),
CONTROL-SIGNAL-EV( $Time, t$ ) $\equiv$ ( CONTROL_SIGNAL = ON )/( $Time, t$ ).

The above events are *software controlled* in the sense that they are under control of a computer based controller of the gas burner. The condition states that the hazard occurs if the ignition is off before the CONTROL-SIGNAL-EV event is issued and the CONTROL-SIGNAL-EV event starts before or coincides with the OPEN-VALVE-SIGNAL-EV event, and all three events overlap for the time period longer than $t_G + t_Z$. In order to prevent the hazard occurrence it is enough to negate the enabling condition associated with the cut set. As the contributing events are software controlled, it is enough to include the negation of the condition to the set of software requirements and to make sure that this requirement is always fulfilled by the software.

## 6.1. Algorithm for calculation of an enabling condition of two events

The enabling condition of a minimal cut set can be derived from the fault tree using formal reasoning. We have developed an algorithm and a tool to perform this analysis. The algorithm works as follows. First, for each gate, we convert the corresponding formal formula in such a way that it is represented as a conjunction of simple relations referring to the *start*() and *end*() functions. For example, the formula (*) can be converted as follows:

$start(e1) + t_G < end(e1) \land start(e1) + t_G < end(e2) \land$
$start(e2) + t_G < end(e1) \land start(e2) + t_G < end(e2)$

For simplicity, let us assume that each gate of the tree has at most two input events. The algorithm works in the following steps:

i.  If events E' and E'' are linked to a common gate then *enabling-condition*(e',e'') is derived directly from the formal definition of the gate. This condition defines the weakest relationship in time between e' and e'', admitted by the definition of the gate.

ii. If E',E'' are linked to different gates and there is an event E''' such that *enabling-condition*(e',e''') and *enabling-condition*(e'',e''') are known, then *enabling-condition*(e',e'') must fulfil the following:

$$\forall\ e' \in \phi(E'),\ e'' \in \phi(E'')\ \cdot$$
$$(\ \exists\ e''' \in \phi(E''')\ \cdot$$
$$\textit{enabling-condition}(\ e',\ e''\ ) \Leftrightarrow$$
$$\textit{enabling-condition}(\ e',\ e'''\ ) \wedge$$
$$\textit{enabling-condition}(\ e'',\ e'''\ )$$

iii. If E' and E'' are two arbitrary events of a fault tree then *enabling-condition*(e',e'') is identified by the following substeps:
  - identify a minimal chain of events through the fault tree which connects E' and E'' in such a way that each two adjacent events in the chain belong to a common gate (for example, if in Fig.2. E'=E7 and E''=E12 then the chain would be E7,E4,E3,E1,E2,E9,E10,E12),
  - identify the enabling condition for each pair of adjacent events in the chain (apply step (i)),
  - identify the enabling condition for E' and E'' by repeated application of step (ii).

The above algorithm can be used to identify the enabling condition for any two events of a given minimal cut set. The crucial step in the algorithm is (ii) where we derive the enabling condition of two arbitrary events E', E'' from the enabling conditions of events E', E''' and E'', E'''. In order to do this we represent the known enabling conditions as directed weighted graph (in Fig.3. we can see such a graph for the enabling condition (*)).

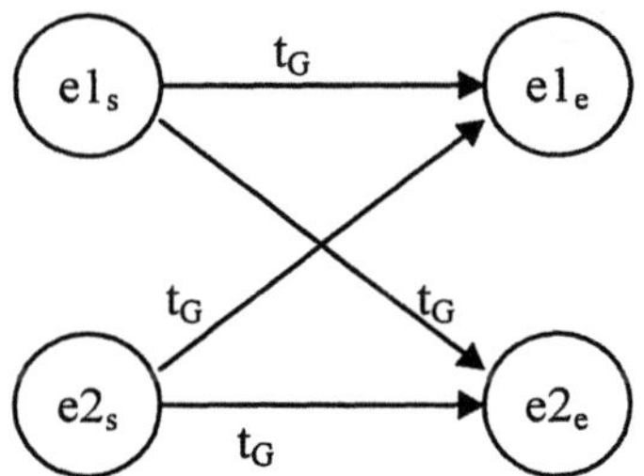


Fig 3. Graph representing the enabling condition (*).

The graph represents time dependencies between the *start* and *end* transitions of the participating events. The two graphs, corresponding to the enabling conditions of E',E''' and E'',E''', are then converted into one (through the matching nodes) and we calculate the weights for all edges which are not represented explicitly. Then the minimal paths connecting the transitions of the events E' and E'' are identified. The time weights on those paths determine the enabling condition for E' and E''.

## 8. Petri Nets Analysis

In [GMW95] it has been shown how a formal definition of a Fault Tree can be used as the specification of an executable model of the tree, expressed in terms of Time Petri Nets (TPN). The net is then analysed with respect to reachability of the state which corresponds to the hazardous event. The analysis provides a precise answer if the hazard can actually occur. The approach assumes that each (formally specified)

240

gate of the tree is systematically converted into its TPN equivalent. Then, through the reachability analysis of the state classes of the resulting net we can verify if, starting from the initial marking of the net, a state which includes in its marking the place representing the hazard is reachable.

In order to apply TPN analysis to Fault Trees we have to convert a given tree into its TPN representation. The CSDM based definition of a fault tree gate can serve as the specification of the TPN construct which simulates the gate behaviour. We have developed  the *CSDM to TPN dictionary* which, for each type of gate, gives its CSDM specification and the corresponding TPN structure. An example entry of  the dictionary is given below.

**Generalisation OR:**

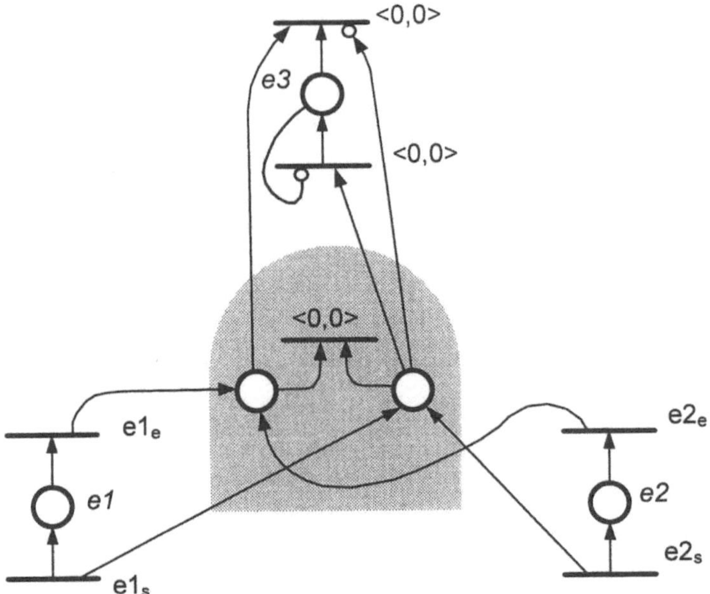


Fig. 4. Petri Net for generalisation OR gate

where static time intervals associated with the transitions of the net are specified as follows:

$$SI(e1_s) = <0,0>,$$

$$SI(e1_e) = <t_{xmin}, t_{xmax}>,$$

$$SI(e2_s) = <0,0>,$$

$$SI(e2_e) = <t_{ymin}, t_{ymax}>,$$

$$SI(e3_s) = <0,0>,$$

$$SI(e3_e) = <0,0>.$$

The CSDM-to-TPN dictionary can be used for systematic conversion of a (formally specified) Fault Tree into its TPN counterpart. Then, through the reachability analysis of the state classes we can verify if, starting from the initial marking of the net, a state class which includes in its marking the place representing the hazard is reachable.

The minimal cut sets analysis extended with time provides for identification of enabling conditions associated with each minimal cut set in isolation. In case when such a condition is maintained *False* the hazard occurrence resulting from the occurrences of the events belonging to the cut set is excluded. However, we should additionally verify if the hazard is still unreachable as the result of the interference of the events from different cut sets (even if each of the corresponding enabling conditions is maintained *False*). To illustrate the problem let us consider a part of the gas burner Fault Tree, as shown in Fig.5.

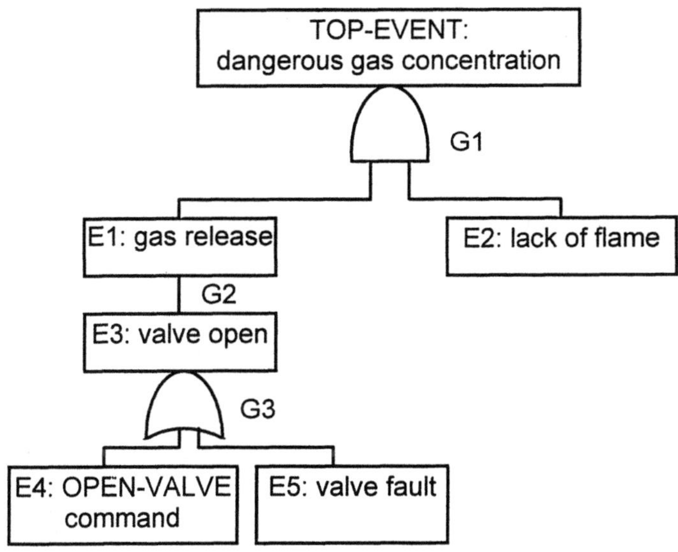


Fig.5. A partial Fault Tree of the gas burner.

Let us consider the following scenario of events. An OPEN-VALVE command is issued by software (E4 begins). Before the time $t_G$ elapses, the software issues the closing command (E4 terminates). However, the valve fails while being closed (E5 starts) at the same moment. In this case, the valve remains open despite the end of event E4, as is illustrated in Fig.6.

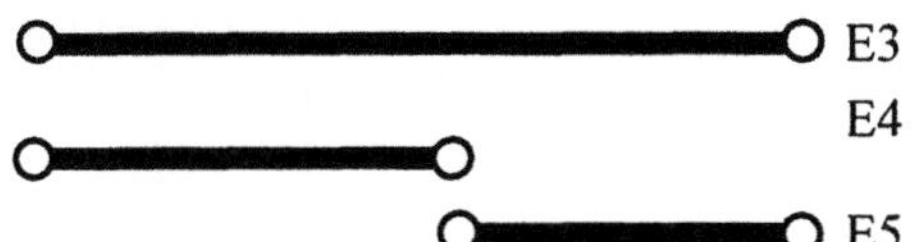


Fig. 6. Time diagram for dangerous scenario

Even if the duration of E5 is shorter then $t_G$, the whole duration of E3 can be longer that $t_G$ because of the concatenation of the events E4 and E5. This problem can not be dealt with by the minimal cut set analysis as the two events (E4 and E5) will not belong to a common cut set (they are the input to an OR gate).

The above problem can be easily analysed with the help of TPNs. The net corresponding to gates G1 and G2 of Fig.5. is shown in Fig.7.

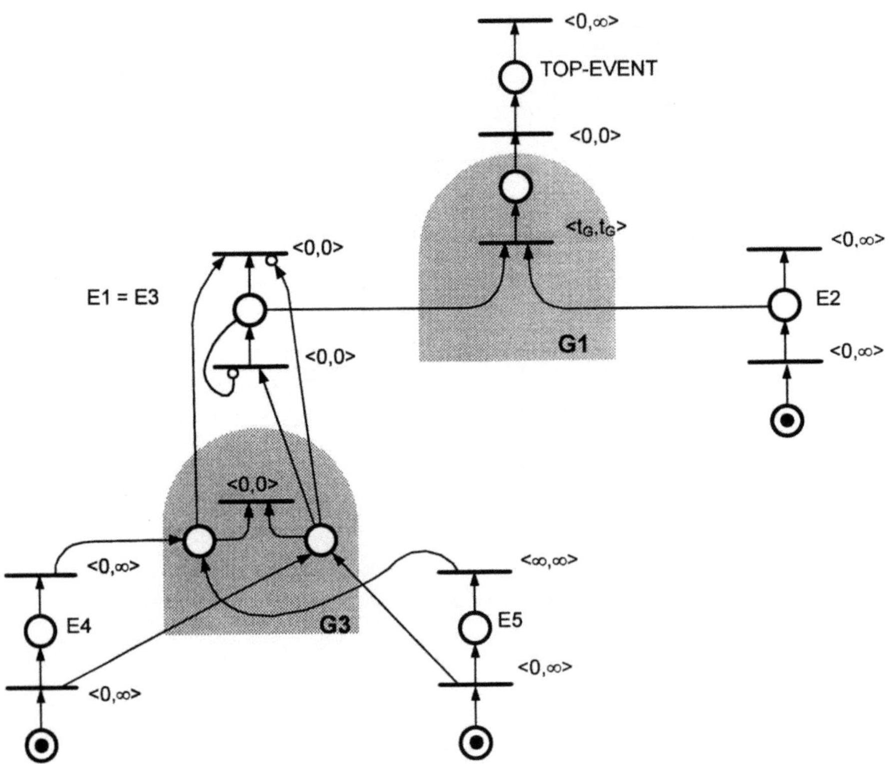


Fig. 7. Time Petri net model of gates G1, G3.

Such a model is then augmented with the negated enabling conditions corresponding to the minimal cut-sets of the tree. For instance, the negated enabling condition of the minimal cut-set {E2,E5} is shown in Fig.8. The negated enabling condition for the minimal cut set {E2,E4} can be added in a similar way.

The reachability analysis of the resulting net reveals that hazard is still possible, although the enabling conditions of the minimal cut sets are maintained *False*.

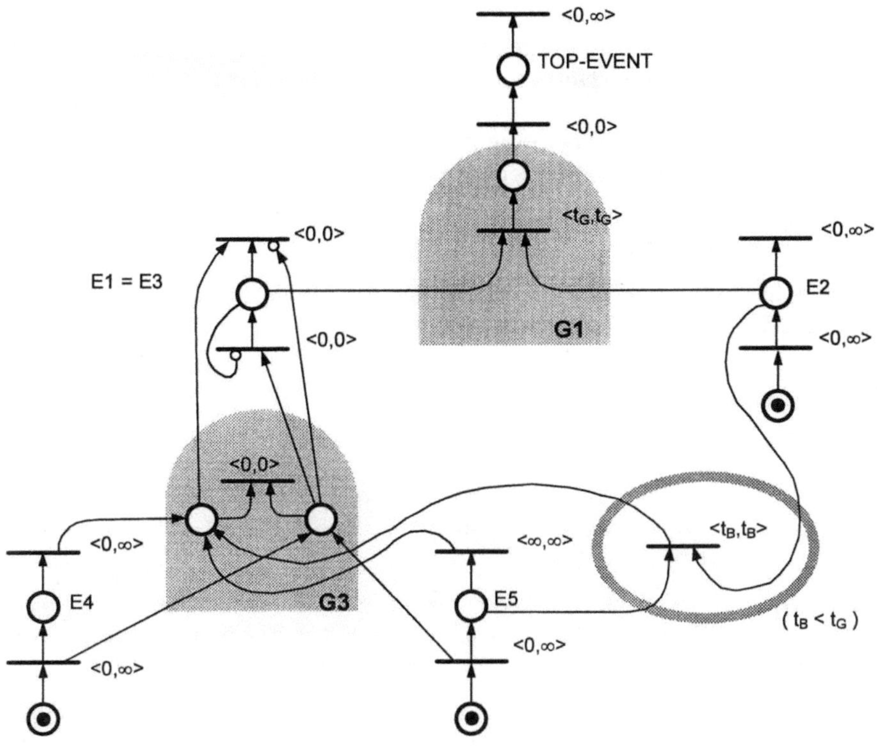


Fig. 8. Time Petri Net extended with the negated enabling condition of {E2,E5}.

The reachability analysis of the resulting net reveals that hazard is still possible, although the enabling conditions of the minimal cut sets are maintained *False*.

# Conclusions

In the paper we have presented a new approach which aims at analysis of timing properties of Fault Trees. The approach takes the conventional Fault Tree as an input and then, through the subsequent steps of formalisation provides for precise interpretation and analysis of the meaning of the tree. In our opinion this is an advantage of our approach which builds on the top of the results of a widely used safety analysis method and adds to it an additional analytical power.

Not all applications will require the analysis of the type presented in this paper. However, there are some systems where hazard clearly depends on the time relationships between the involved events. A small case study presented in this paper illustrates such a situation.

In practical terms, we consider that our method can be embedded into a set of tools which would support its application by a safety engineer. He would take a conventional Fault Tree as an input and then extend it with a formal definition and perform the timing analysis by invoking the corresponding algorithms. We have already developed prototype tools for some of those algorithms.

The method is still under further examination. Presently, we are experimenting with it with respect to some case studies derived from the nuclear industry [WČ96].

## Acknowledgement

This work has been partially supported by the ISAT Project of the European Union COPERNICUS Research Programme (JRP 1594).

## References

[BCG90]     R. E. Bloomfield, J. H. Cheng, J. Górski, *Safety Analysis: a feasibility study into the develpment of a generic Safety Description Method (SDM)*, EUREKA Project SEW 263, Adelard, 1990 (unpublished)

[Górski94]     J. Górski, *Extending Safety Analysis Techniques With Formal Semantics*, In Technology and Assessment of Safety Critical Systems, (Edited by F.J. Redmill), Springer-Verlag, 1994

[GMW95]     J. Górski, J. Magott, A. Wardziński, *Modeling Fault Trees Using Petri Nets*, In Proceedings of the 14th International Conference on Computer Safety, Reliability and Security SAFECOMP'95 (Edited by G. Rabe), Springer-Verlag, 1995

[GW95]     J. Górski, A. Wardziński, *Formalising Fault Trees*, In Proceedings of the Safety Critical Systems Symposium (Edited by F. J. Redmill and T. Anderson), Springer-Verlag, 1995

[GW96]     J. Górski, A. Wardziński, *Deriving Real-Time Requirements for Software from Safety Analysis*, In Proceeings of the 8th EUROMICRO Workshop on Real-Time Systems, IEEE Computer Society Press, 1996

[Jones90]     C. B. Jones, *Systematic Software Development using VDM*, Prentice Hall Int., 1990

[WČ96]     A. Wardziński, M. Čepin, *On Integration of Probabilistic and Deterministic Safety Analysis*, 3rd Meeting Nuclear Energy in Central Europe, Portoroz (Slovenia), September 16-19, 1996

# SACRES - Formalism for Real Projects


Alvery Grazebrook

i-Logix UK Ltd., 1 Cornbrash Park, Bumpers Way, Chippenham,
Wilts SN14 6RA. Tel. (01249) 446 448


## Abstract


The SACRES project has one goal in mind - the production of extremely reliable systems. The project brings together a powerful tool-set and supporting development methods. The tool-set is integrated through the DC+ format, which is a synchronous language, and embeds the specification notations. The methodology is developed using the experience of world-leaders in industry.

The tool-set will be composed of the specification notations, a verification tool-set, an automatic code-generator and code verification. The specification notations comprise Statecharts, Signal and Symbolic Timing Diagrams. The verification tool-set supports the automated proof of properties. The code-generator is tightly linked to the code-verification, which proves the correctness of the generated code against the source format.

In this presentation, I will discuss the architecture of the SACRES tool-set, the approach to formal verification, and look to the results that we expect to achieve in the project's future.


# 1. Overview

The SACRES project is intended to provide tools and supporting techniques to the designers of embedded control systems. In particular, tool support is required for safety critical embedded systems. In the UK market the partners are motivated, in part, by new standards for the certification of safety-critical systems such as Def-stan 00-55. There are similar pressures for improving the evidence presented to various certification authorities in other European countries.

The project involves the integration of existing tools, the development of new technologies, and the development and trial of supporting development methods.

The existing tools are Statemate (from i-Logix), SILDEX (from TNI), Timing Diagrams and a formal verification tool-set (with components from Siemens, Siemens Nixdorf and OFFIS). The Statemate and SILDEX specification tools are integrated by providing interfacing information in a System Specification Language (SSL), and translating the composite model into the DC+ format. DC+ is a synchronous specification language, and was developed in the Synchron project. [MAR95] [Lan95] [BEN94] [tem95] [Les91]

The new technologies under development include:
- A code-generator producing sequential code (C or Ada,) for targeting specific real-time OS or distributed microprocessor architectures.
- A code-validation tool, using formal verification to automatically prove the correctness of the generated code.
- Formal verification of properties, improving upon technology developed under the FORMAT project.[DAM94] [HEL93] [SCH93]

The industrial partners will each use the tool-set to develop a system, and the project will use the experience gained to write a handbook on how best to apply the tool-set. The industrial partners are:
- British Aerospace, Military Aircraft
- Siemens
- SNECMA

Each of the partners are developing a system that is typical of their usual work. For example, BAe is developing an avionics system, and Siemens is developing systems for industrial process control.

## 2. Architecture

The SACRES tool-set consists of four major functional components, exchanging information through the DC+ format. The components are:
- Specification front end - for creating and editing the specification
- Verification tool - for verifying the correctness of the specification
- Code-generation tool - to generate C or Ada or distributed code
- Code validation tool - to prove the correctness of C and Ada code.

The user's access to the tools is through a command-level interface.

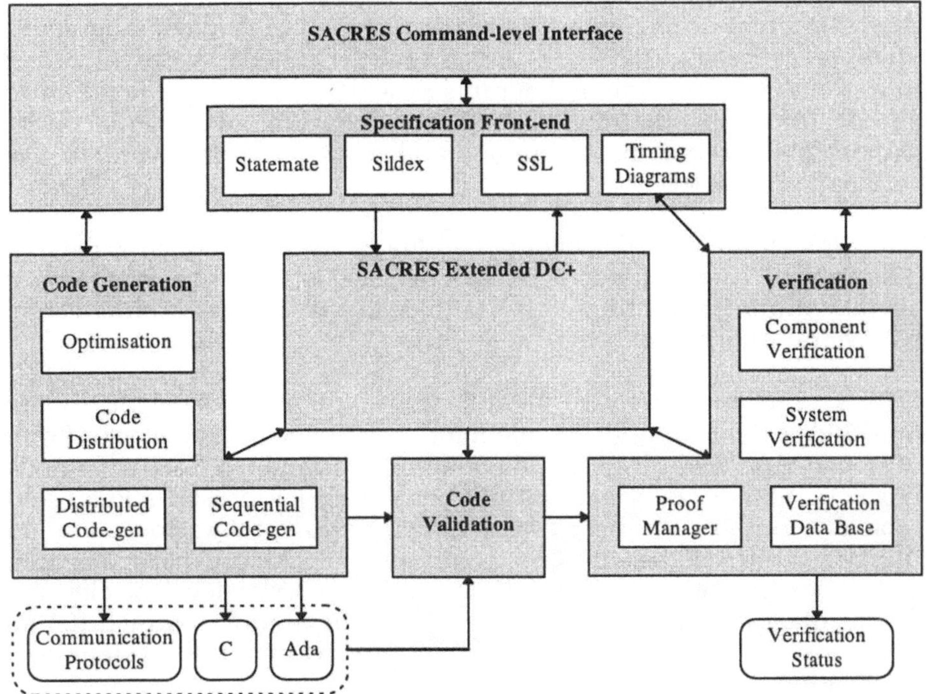

Fig. 1: SACRES Global Architecture

The Global Architecture diagram shows the tools, and the general flow of information that connects the tools. It is worth noting that the different tools do store information locally as well as reading model data from DC+. For example, the specification tools each have their own file structures. Translators are being developed within the project to convert tool information to and from DC+.

## 2.1 Specification Tools

It is difficult to understand the SACRES architecture without a knowledge of the tools that already exist and the knowledge and technology from previous projects.

The Statemate[ILOGIX] [HAR90] and Sildex[TNI] tools are both established in the embedded systems market. Both tools use a structured notation, using functional blocks interconnected with data-flows or signals to define the relationship between components in a model. Statemate uses Statecharts (a hierarchical state-machine notation) combined with a textual mini-spec. notation to describe system behaviour. This is particularly appropriate for describing moding and sequences of operation. Sildex uses Signal and Grafcet to build a data-flow model of a system. This is particularly appropriate for modeling signal processing and control-law applications.

Both Statemate and Sildex are operational specification languages. That is to say, they are executable notations with a direct interpretation. The Timing diagrams provide a declarative notation to specify system operation. They are typically used to specify constraints on system execution, and only have relevance to the verification tool-set.

The SSL (System Specification Language) is used to describe the hierarchical composition of behavioral blocks, and the mapping of signals (data-flows) between the blocks. It doesn't contain any behavioral information.

SSL is used exclusively to identify the mapping between Statemate and Sildex models, and the mapping from behavioural blocks (from Statemate or Sildex) to timing diagram notations. Each leaf-node block in SSL is allowed multiple behavioural representations, and typically contains multiple timing diagrams to specify constraints on the behaviour. Higher-level, or composite, blocks are used to group leaf-node components. These blocks refer to lower level blocks and timing diagrams for the composite block. The ability to attach timing diagrams to composite blocks is important for system-level verification, discussed in the next section.

The specification tools cover both operational and declarative methods of modeling system behaviour. They also provide methods for modeling continuous and discrete systems. The SSL provides a means for constructing larger systems out of components defined in the other tools. It is my hope, later in the project, the tool-set will hide the details of SSL from the specification author, and to make better use of the structuring capabilities that are already available in Statemate and Sildex.

The combination of graphical notations in this project should provide a mechanism for formally specifying a large range of embedded systems. The resulting specifications should be easy to read, both by the specification authors, but also by end-users and managers, who don't necessarily share their specialist skills.

## 2.2 Integration Through DC+

The combined specification is entered in the individual tools, but to allow the SACRES toolset to act as a coherent toolset, it is vital that all the tools work with a coherent, single system model.

This model is stored in the DC+ format. The characteristics of DC+ are central to the operation of the verification and code-generation tools. It is possible to specify both operational and declarative expressions in the format, so the expression of proof constraints are possible alongside the expression of behaviour. This capability is used by the verification tools, and may also be used in the code-generation tools.

DC+ also has several distinct "layers". In its loosest sense, a DC+ model describes functions that relate inputs to outputs using the internal state, where the application of the function is based on the occurrence of abstract clocks. That is to say, the outputs of a function exist when the inputs all exist at the same time.

One of the refinements of DC+ uses a technique called clock expansion to synchronize the different functions. This is necessary for proof, when relating the occurrence of different outputs.

Another refinement allows scheduling of functions. Generally, in DC+, all functions are assumed to be operating in parallel (where this is relevant.) The dependencies between functions are based on the relationship of the inputs of one function with the outputs of another. In order to generate code for a sequential language (like C and Ada) it is necessary to pre-schedule the functions, so that they will be executed in the correct sequence.

## 2.3 Verification Tool-set

The verification tool-set is mostly based on technology assembled in the FORMAT project. The FORMAT project provided an integration between a BDD based model checker, behavioral information from VHDL models, a textual state-machine representation, and timing diagrams. Supporting tools included a proof manager and verification database.

The proof checker can be used in two modes. It is either treated as a tautology checker, which provides a means to check consistency between separate declarative specifications (e.g. comparing assemblies of timing diagrams) or alternatively to verify the consistency of operational behaviour with a declarative specification. In the FORMAT project this was used to check that VHDL blocks satisfy the constraints defined in a set of timing diagrams.

In the SACRES project, the component verification is used to check that the behaviour of components identified in SSL are consistent with the constraints set upon them in timing diagrams. The component behaviour can include expressions in Statemate or Sildex, or a combination of both.

Component verification has been used by Siemens[SNI] since 1990. Models as large as 1000 pages of SDL spec have been checked, as well as similarly large VHDL models. The technology for verifying VHDL is commercially available from AHL.[AHL]

For example, the state-machine for the indicator on a car might specify the flashing rate by defining a target duration for the indicator to be on (On_Time) and a target time for the indicator to be off (Off_Time). The Statechart would look like this:

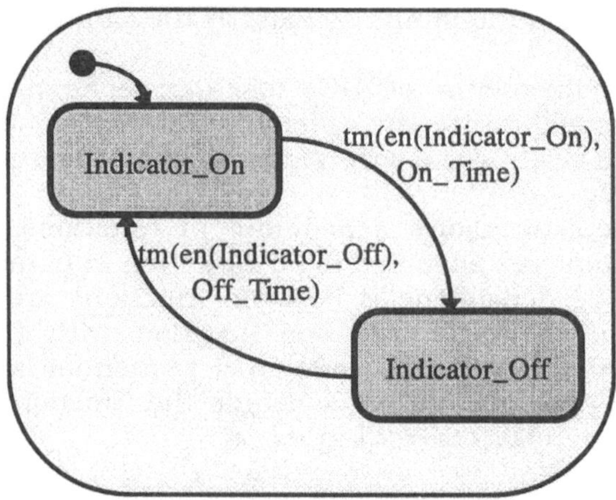


Suppose that the specification also defines that the flashing interval must be within certain (Government defined) limits. This would be expressed in a timing diagram (declarative specification). For example,

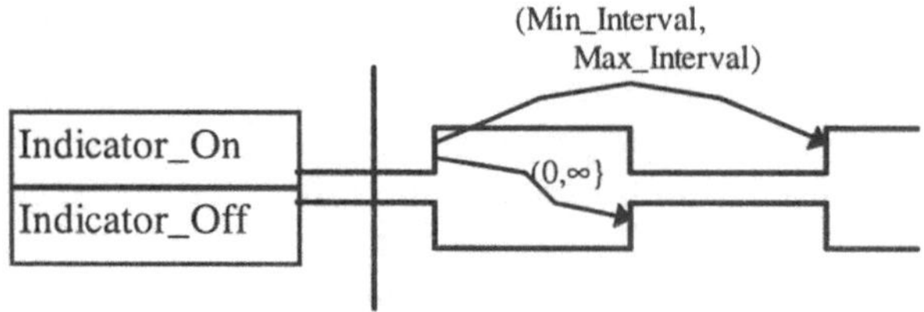


The component verification tool would then be used to check that the state-machine did in fact implement a machine that would always meet the constraints.

At present, the component verification is only able to handle time relationships with zero or infinite time spans - equivalent to defining ordering of events and eventual completion (non-deadlock) of events. Within this project, the component verification technology will be extended to support an explicit representation of time.

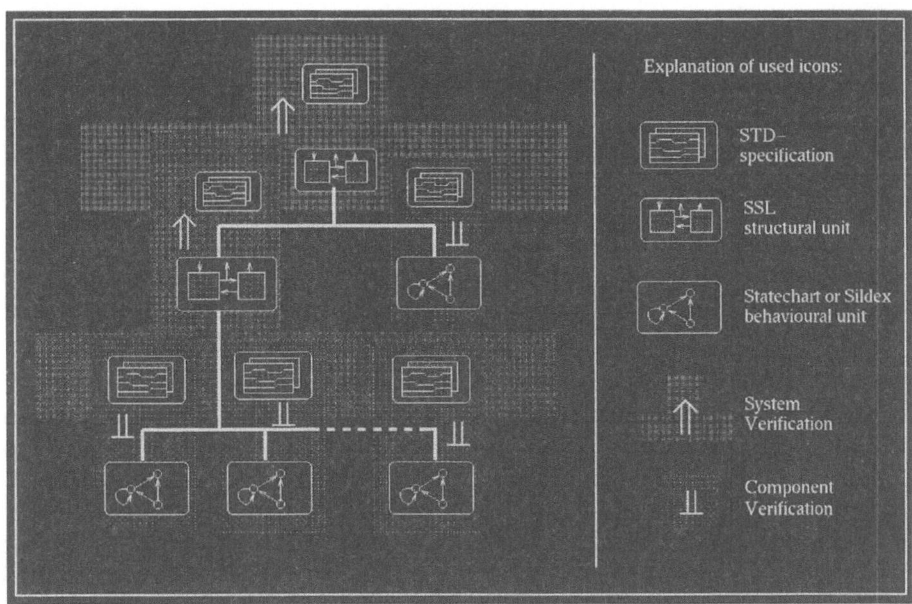


Component verification, like all the current automatic verification technologies is limited in the complexity of models that can be verified. Siemens, Siemens Nixdorf and OFFIS[OFFIS] are working on system verification as part of the project.

System verification uses a combination of tautology checking and the structuring information from SSL to verify consistency between properties already proven for a component, and properties defined for a higher level in the hierarchy.

The combination of component verification and system level verification provides a powerful way to check properties of large systems.

## 2.4  Code Generation

Code-generation capabilities already exist for DC+ and the operational specification tools - Statemate and Sildex. These code-generators are all independent, and are targeted for single-processor systems.

As can be seen from the Global Architecture diagram, the SACRES project includes capabilities for distributed code-generation, including the generation of code to use the communications infrastructure available on the target system. This work is being performed by TNI and INRIA[INR].

A single target code-generator forms the base for a distributed code-generator, by generating the code that will execute on an individual task / processor. In addition to this, a distributed code-generator must contain a mechanism for partitioning  behaviour across the different targets, and generate the additional code for communicating between them.

It is necessary to provide a significant amount of information about a target architecture in order to generate correct code for the architecture.

## 2.5  Certification

In safety critical systems, it is not enough to have a code-generator to produce efficient code from the model. It is also necessary to demonstrate that the generated code is a correct implementation of the model.

At the outset of the project, we considered three alternative strategies for solving this problem. One strategy involved the automatic generation of test-cases, and the execution of those test-cases against the generated code; another, a proof of correctness of the code-generator itself; lastly to provide an automatic mechanism to prove that the generated code was consistent with the DC+ model.

The project chose to follow the final mechanism, to prove that the generated code was consistent with the DC+ model. Some partners considered the test-case mechanism inadequate, because it is impossible to cover all possible executions of any significant model using this technique. The proof of correctness of the code-generator was considered too difficult, and also would fix the implementation of the code-generator, requiring complete re-proof of the code-generator if any algorithms were changed or improved.

The certification approach uses a novel technique for proving the correctness of the generated code, based on research at the Weizmann institute[WEI].

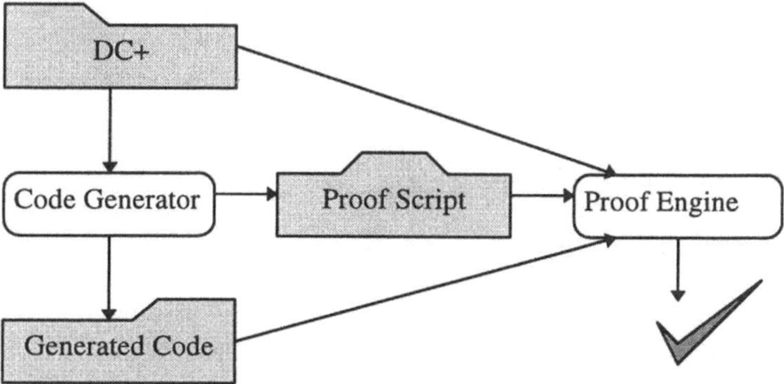



In addition to generating code, the code-generator also writes a proof script to prove the individual transformations that are used to translate the DC+ model into the code. This script drives the proof tool to check the correctness of the generated code against the DC+ model. The proof engine would therefore show the code as erroneous either if the code itself was incorrect or if the proof script was incorrectly generated.

The advantage of this approach is that the "intelligence" in the proof mechanism is contained in the code-generator. This means that changes only need to be made in one place - if you alter the code-generation algorithms, you alter the proof script generation algorithms at the same time. It also means that the proof engine itself can be extremely simple, and it should be easy to demonstrate that a simple proof engine is correct.

## 3. Industrial Involvement

The involvement of the industrial partners, BAe, Siemens and SNECMA, is to prove the value of the tool-set on projects. Confidentiality prevents me from describing the actual projects, but in all cases, the objectives of the partners with respect to the SACRES project are the same:
- To ensure that the tools work on an industrial scale problem
- To provide guidance to the people who are building the tools on the way in which the tools are likely to be applied.
- To adapt their development practices to take advantage of the new capabilities offered by the tool-set.

They are also working on a common set of guidelines for using the tools - in effect a development methodology suited to the use of the SACRES toolset. These guidelines will be published towards the end of the project.

# 4. Progress

The project started with a "baseline" phase, in which the partners wrote the requirements for the tool-set, agreed on file-formats and other interfacing issues, and planned the future developments. This phase was completed in April '96.

The prototype building phase will complete soon after this paper is presented, in March '97. During this time, all the components of the toolset will be interfaced to DC+, and some improvements will be made to existing tools. Also a preliminary version of the code certification will be built.

The project continues with an evaluation phase, in which the industrial partners exercise the tool-set on their own projects, followed with a further improvement phase in which the tools are developed to cover shortcomings discovered in the evaluation phase.

The result will be a method, for use by design engineers, that is familiar and easy to learn. The method will support development of nearly every kind of control system, from aircraft avionics to industrial plant control. Most of all, the method will provide good supporting evidence that the design is correct, with formal proof of safety properties, and formal verification of the correctness of the code against the model.

# 5. References


[AHL] Abstract Hardware Limited, 1 Brunel Science Park, Kingston Lane, Uxbridge, Middlesex, UB8 3PQ, UK. Tel. +44 1895 258501  Fax. +44 1895 259437  Web: http://www.ahl.co.uk/

[BEN94] A. Benveniste
Synchronous languages provide Safety in reactive systems design
*Control Engineering*, Sept. 1994, pages 87-89.

[DAM94] W. Damm, B. Josko, R. Schlör
Specification and Verification of VHDL-based system-level hardware designs.
E. Börger, editor, *Specification and validation methods*

[HAR90] D. Harel, H. Lachover, A. Naamad, A. Pnuelli, M. Politi, R. Sherman, A. Shtull-Trauring, M. Trakhtenbrot
STATEMATE: A Working Environment for the Development of Complex Reactive Systems.
*IEEE Transactions on Software Engineering.* Vol. 16, No. 4, April 1990.

[HEL93] J. Helbig, R. Schlör, W. Damm, G. Döhmen, P. Kelb
VHDL/S - integrating Statecharts, timing diagrams and VHDL
*Microprocessing and Microprogramming 38*, pages 571 - 580, 1993

[ILOGIX] i-Logix UK Ltd., 1 Cornbrash Park, Bumpers Way, Chippenham, Wilts, SN14 6RA, UK. Tel. +44 1249 446 448  Fax: +44 1249 44 73 73  Web: http://www.ilogix.com/

[INR] Dr. Albert Benveniste,  Institut de Recherche en Informatique et Systèmes Aléatoires, Campus universitaire de Beaulieu, 35042 Rennes, Cedex, France  Tel +33 299 84 71 00  Fax +33 299 84 71 71

[Lan95] (author unknown)
Languages synchrones : l'essai a transformer
*01 Informatique*, 25 Aug. 1995, page 42.

[Les91] (author unknown)
Les systemes reactifs ou la modelisation des refrexes
*Micro-systemes*, Oct 1991, Pages 178-181.

[MAR95] P. Marsier, C. Medigue.
Linear and non-linear analyse *(sic)* of heart rate variability : a minireview in Cardiovascular research.
*Journal of the european society of cardiology*, Oct 1995, pages 371-379

[OFFIS] Prof. Dr. Werner Damm, OFFIS, Carl v. Ossietzki Universitaet Oldenburg, Technische Informatik, FB Informatik, Postfasch 2503, 26111 Oldenburg  Tel. +49 441 798 4502  Fax. +49 441 798 2145

[SCH93] R. Schlör, W. Damm
Specification and verification of system-level hardware designs using timing diagrams.
*Proceedings, The European Conference on Design Automation with the European Event in ASIC Design*, pages 518-524, Paris, France, Feb. 1993. IEEE Computer Society Press.

[SNI] Siemens Nixdorf Informationssysteme AG, System Unit BS2000, Product Planning, Otto-Hahn-Ring 6, D-81739 München, Germany.  Tel. +49 89 636 44049  Fax +49 89 636 44352

[tem95] (author unknown)
temps reel dur : Accord sur les languages synchrones
*01 Informatique*, 24 Feb. 1995, page 22.

[TNI] Techniques Nouvelle d'Informatique, case postale no1, 29608 Brest Cedex, France. Tel. +33 298 05 27 44  Fax. +33 298 05 63 50.

[WEI] Prof. Amir Pnuelli, Weizmann Institute of Science, Dept. Applied Mathematics and Computer Science, Rehovot, 76100, Israel. Tel +972 89 34 34  Fax +972 89 344 122

# Product Monitoring for Integrity and Safety Enhancement


Robin Whitty
Centre for Systems and Software Engineering
School of Computing, Information Systems and Mathematics
South Bank University
London SE1 0AA, UK


### Abstract


We report on an attempt to introduce software product metrics into the process of assuring and controlling the quality of safety-critical software. Ironically, a number of factors which seemed to auger well for this attempt eventually militated against its success.


## 1 Introduction

The PROMISE project (Product Monitoring for Integrity and Safety Enhancement) attempted to deploy methods of software product measurement in the safety critical domain. The specific focus of the project was the software used in radiotherapy machines and we had the co-operation of Philips Medical Systems-Radiotherapy (PMS-R) who allowed us to examine the products and processes connected with a particular machine, the SL series. We aimed to establish how product measurement might be integrated with the quality assurance and control procedures of PMS-R and of software developers generally within the safety-critical domain. We think the project failed in this aim (although it gave value for money in unexpected ways, as must often be the case with speculative research). We shall try, in this paper, to see what lessons can be learned from this failure.

The area of 'software metrics' began life as the study of product measurement. During the 1970's a number of algorithms were invented for capturing numerically attributes of source code impacting upon its reliability, ease of maintenance, ease of comprehension etc. For example, the classic paper of McCabe [McCabe 76] proposed a measure of testability (cyclomatic complexity—the size of a certain path set in the program graph) and a measure of structuredness (cyclomatic complexity shrunk in proportion to how well-structured the code is) and embraced both under the general heading of 'complexity measurement'. The idea was that testability, structuredness and complexity (attributes of the product) were drivers of the cost of testing and maintenance (attributes of the process).

Since the 1970's it has come to be realised that such simple algorithms as McCabe's cannot be adequate to describe attributes such as testability, structuredness and complexity, and that there is, in any case, no simple relationship between such code attributes and attributes relating to process cost or effectiveness, however much such relationships may appeal to our intuition It now seems naive to think that a 'physical' attribute of software might explain its behaviour during use or maintenance in the way that the gravitational attraction of the planets explains their behaviour in space.

Nevertheless, a corpus of pragmatic techniques has built up for using product measures without relying on their accurately describing attributes or without having a knowledge of their precise relationships with process measures. These techniques include Fuchs' testing strategy, product profiling, and outlier and regression analysis. The PROMISE and DATUM projects added a new technique by applying Bayesian networks to the problem. In the right circumstances, these techniques may yield information which is useful to software engineering or managers. The hypothesis of the PROMISE project was that the safety-critical domain provided the right circumstances. I think we have refuted this hypothesis; a number of factors which seemed initially to bode well for its confirmation eventually militated against it.

In the next section we shall describe in more detail the uses mentioned above to which product measurement can be put. Section 3 describes the PROMISE project, giving details of how we tried to make the project a success and why, in our opinion, it eventually failed. Section 4 then revisits the subject of putting product measurement to use in the light of the lessons learnt from PROMISE.

The opinions expressed here are those of the author and do not necessarily reflect the views held by PMS-R or by the project sponsors.

## 2 The Appropriate Use of Product Measurement

### 2.1 The Limitations of Product Measures

Software product measures are any numbers which can be calculated from some output of the software development process. In theory this can include specification documents, test data, user documentation or object code, and the numbers can be obtained by static or dynamic analysis. However, the focus of the subject has always been static measurement of source code and that will be our focus. The archetypes are the measures of Halstead [Halstead 77] and McCabe [McCabe 76], the former based on the idea that complexity in code is determined by the number and variety of its operators and operands, the latter trying to capture testability in terms of a path count, as described above. Not much has changed in the way that product measures are devised or used, although the sophistication of the ideas has advanced considerably. All this is very comprehensively described in [Fenton 96].

A distinction is sometimes made between a code 'metric' and a code 'measure' in order to emphasise that a measure is trying to capture a specific attribute. The representational theory of measurement decrees that if a metric is to *measure* an attribute then it must preserve numerically the ordering which the attribute puts on the entities concerned. So whenever program A can be certified as being more testable than program B, its cyclomatic number must be greater than that of program B, otherwise whatever cyclomatic number might be measuring, it is not measuring testability. This line of argument effectively kills any pretensions which a product measure may have to capturing any intuitively important attributes of code [Fenton 92]. When we say that a path count is a measure of testability we mean this only in a very informal and inadequate sense.

The question of whether or not a particular number 'describes' testability may be dismissed as a question of semantics. We may as well say that our number in some sense embodies some ill-understood notion of product 'complexity' and proceed without further introspection to identify its relationship with what really matters, the behaviour of the product in some process—use or maintenance or testing or whatever. Such attempts seem doomed to failure; with the possible exception of performance measurement, no interesting 'physical' law has ever been established relating product to process in the software domain. Empirical support for such relationships may easily be demolished [Hamer 82, Shepperd 88, Courtney 93].

We may not be able to place much faith in product metrics as measures of important attributes, or to expect 'physical' laws to relate them to process attributes of software involving time or cost. But there is still something to be rescued from the situation; a number of techniques have been devised for putting product measures to some practical use. These techniques will now be briefly described; together with the DATUM/PROMISE application of Bayesian networks, which will be briefly described in section 3, they represent the options which were available to the PROMISE project.

## 2.2 Pragmatic Use of Product Measures

To make things more concrete we shall invent a toy scenario. A software system contains eight software modules $M_1$, ..., $M_8$. Testing reveals the following number of faults in the modules:

| Module | $M_1$ | $M_2$ | $M_3$ | $M_4$ | $M_5$ | $M_6$ | $M_7$ | $M_8$ |
|--------|-------|-------|-------|-------|-------|-------|-------|-------|
| Faults | 3 | 21 | 1 | 6 | 18 | 1 | 0 | 17 |

Table 1: fault data for an imaginary collection of modules.

This is our process data: the number of faults found during testing. Suppose we also have McCabe's cyclomatic complexity for each module. We shall suppose the following values have been calculated:

| Module | $M_1$ | $M_2$ | $M_3$ | $M_4$ | $M_5$ | $M_6$ | $M_7$ | $M_8$ |
|--------|-------|-------|-------|-------|-------|-------|-------|-------|
| McCabe | 10 | 31 | 5 | 9 | 4 | 1 | 16 | 12 |

Table 2: complexity values for the modules in Table 1.

We shall now describe four options at our disposal for putting this data to use. The first three are elementary statistical techniques; their use in software engineering generally is much more fully described in [Lockhart 93], [Burr 96] and [Fenton 96] but they are worth reviewing here since they are easy to describe and will make our subsequent discussion concrete and self-contained.

### *Product Profiling*

This is the most unambitious use of product metrics. The process data is ignored and the product data is plotted and used as a picture or profile of the product. For example, the McCabe profile might be plotted as shown in Figure 1 (box plots provide an alternative view of the same data).

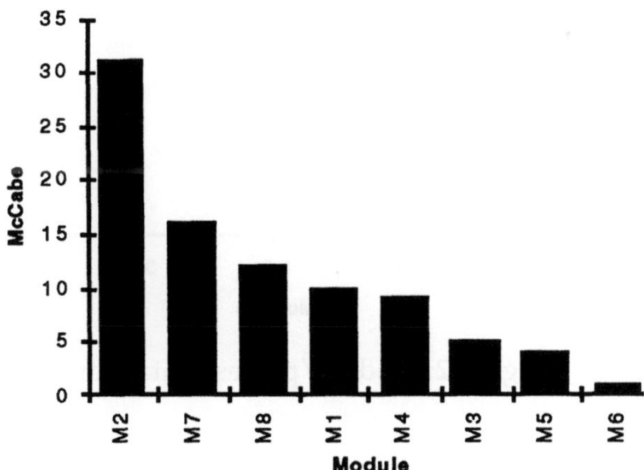


Figure 1: product profile.

Such profiles are clearly of little use on their own, but if they are kept for many systems or for many versions of one system then anomalies or degradations may become apparent and may be acted upon. This use of metrics may, for example, be of use to procurement departments of large organisations which need to monitor the complexity of a bought-in product during the course of many updates or releases of the product. Experts may in time be able to spot the 'fingerprint' of a bad or deteriorating product.

## *Outlier Identification*

At a slightly more ambitious level, we may at least presume some relationship between our product and process data and use deviations from this relationship as data for decision making. For example, suppose the data above are plotted as a scatterplot, as shown in Figure 2.

We may induce a trend from this picture as shown in the sloping line. In effect, we are proposing a model which says that fault rates are proportional to McCabe's cyclomatic complexity. But we are not, at this point, placing enough faith in the model to use it to make predictions. Nevertheless, on the basis of Figure 2 we might decide that the point at (16, 0) represents a module (module $M_7$) which has an unusually low fault count for its 'complexity', based on the trend we have induced. This decision may lead to selection of the module for review.

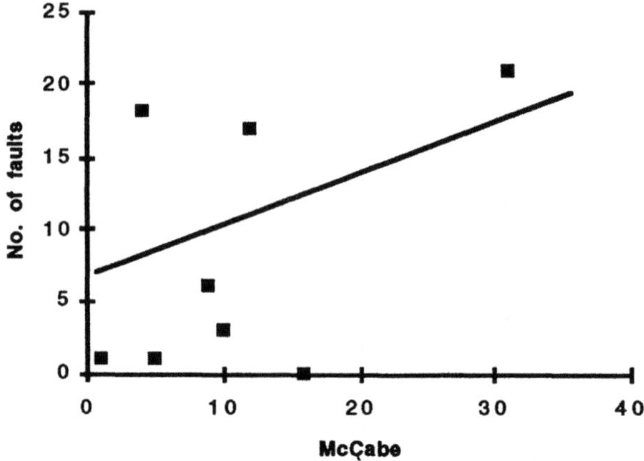


Figure 2: product/process data plot.

## *Regression Analysis*

We  may be more explicit about the fact that we are building a statistical model by giving the product metric the status of an independent variable. The dependent variable is the process measure of interest (in Figure 2, the number of faults). In fact, the best-known example of this approach is Boehm's COCOMO model of software development cost:

$$\text{Person months} = a.\text{KDSI}^b.$$

The value of such models relies heavily upon the amount and quality of process data available. In the software domain this usually presents a grave difficulty: it is very rare to find organisations which systematically collect good quality data on their processes, such as effort to complete a project.

If we do have available process data corresponding to a collection of product measures then there can be no harm in building a statistical model of this correspondence and using it to make predictions. Of course the predictions have to be used with caution. In the case of the PROMISE project, it certainly seemed possible that a model like

$$\text{No. of faults} = \sum a_i \times \text{metric}_i$$

could be constructed with some predictive power. To continue the example above, the trend line in Figure 2 has an equation something like:

$$\text{No. of faults} = 0.3 \times \text{McCabe} + 7.$$

Suppose that module $M_7$, which exhibited zero faults, is reviewed and turns out to have been missed during unit testing. On the basis of the above equation, we might now test this module with the expectation of finding $0.3 \times 16 + 7 \approx 12$ bugs. Regression analysis will allow us to put a figure on our level of expectation. Normally the data is not good enough for this figure to be very high.


*Fuchs' Test Strategy*


What if we have done no testing at all? Surely, then we have no process data and are reduced to profiling? Fuchs has proposed an ingenious method[1] whereby, product measures may still be applied in the testing process. We know, with hindsight, that the numbers of faults which will be found are as shown in Table 1. Suppose we happened to test the modules in the order most-faulty-first. The cumulative total of faults found is represented in the top curve of Figure 3. If we tested in the reverse order, we would instead get the lower curve.

In many cases a deadline will be set after which integration begins. This may well occur before all units are tested. If there was only time to test 75% of the modules then it is clear from Figure 3 that testing in the worst case order will result in many more of the detectable faults going through to integration: nearly 60% instead of less than 2%.

Now suppose that we implicitly assume the same model that was explicit in Figure 2, that fault rates are proportional to McCabe's cyclomatic complexity. According to this model, we should test the modules in decreasing order of McCabe's cyclomatic complexity. In this case we get a middle curve, as shown in Figure 4.

---

[1] This method has been described before [Pulford 95, pp. 164–167] but unfortunately without being attributed to Norbert Fuchs who invented it during the ESPRIT project COSMOS. Lionel Briand has told me that the method has independently been tried at the Software Engineering Laboratory, University of Maryland with some success.

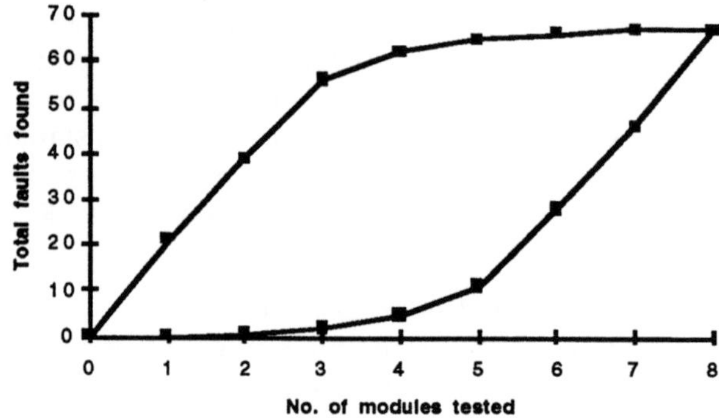

Figure 3: best and worst rates of fault detection.

Unfortunately, since our implicit model is not very accurate we do not achieve the upper curve. Nevertheless, we do a lot better than the lower; at the 75% tested stage, less than 30% of detectable faults are getting through.

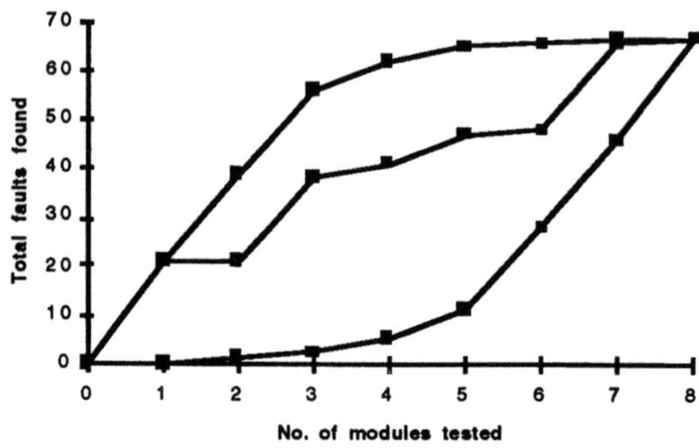

Figure 4: the Fuchs method of testing.

## 2.3 Summary

We have seen that product measurement cannot actually capture product attributes which seem to be important. Even if they could we have seen that there would be no obvious way to build accurate models of their relationship with process measures such as fault-rates, or development or maintenance costs. We have listed four uses of product measures which side-step these limitations: profiling, outlier analysis, regression analysis and the Fuchs method. The first and last of these do not even require any process data, although the Fuchs method assumes an implicit model relating product and process attributes.

The PROMISE and DATUM projects have added one further technique to the list: the application of Bayesian networks, which we shall mention again shortly. As far as we know there are currently no other appropriate ways of using product metrics. Still, we had something to go on for our work with Philips Medical Systems-Radiotherapy; it was certainly not the case that we were seeking to apply, either naively or cynically, a moribund science.

# 3 The PROMISE Project

## 3.1 Background

Philips Medical Systems-Radiotherapy (PMS-R) is one of the major suppliers of radiotherapy machines. They produce a series of machines under the generic name SL, for example SL15, SL18, SL20, SL25 (all four  having dual X-ray and electron energies) and the SL75-5 (X-ray only).  They also manufacture simulators—the SLS23 and SLS38; a treatment planning machine called SLP 3-D; an imaging system—the SRI 100 and lots of other add-ons such as the MLC (multileaf collimator, which provides accurate shaping of the radiation beam). The safety-critical nature of these machines is easily appreciated, especially since the Therac disasters which received much publicity [Leveson 93] and, indeed, represent one of the few recorded sources of genuinely software-attributable deaths [MacKenzie 95].

South Bank University had had some involvement with PMS-R for a couple of years prior to  the commencement of the PROMISE project, taking the form of two MSc student placements on their site at Crawley, Sussex, and a further student providing technical support from South Bank. The final MSc project [Nakassis 92] involved installing the QUALMS software metrics tool at PMS-R and trialling it on some of PMS-R's PL/M-86 code (the principal language used). The project was successful, generating some interesting results and getting a write-up in the in-house quality assurance newsletter. Subsequently, PMS-R moved from a centralised mini-computer architecture to a PC-based client-server architecture. This made some of the previous work with QUALMS redundant.

PMS-R systematically collect fault data which can in many cases be used to trace faults to particular code modules. They have rigorous version control, meaning that the evolution of code can be reconstructed. They already had some experience in using software metrics, as well as other quality procedures such as inspections. They therefore seemed to be ideally suited for deploying the techniques described in the previous section.

## 3.2 Achievements

The plan of the project was, briefly, to reinstall QUALMS on the PC architecture at PMS-R; to analyse the likely impact of the techniques described earlier using existing fault data and code (reconstructed through versioning if necessary); and to then install some product metrics-based quality assurance and control facilities for use on future development and maintenance projects.

Towards this end members of the PROMISE team carried out the following tasks:

### Testability Metrics

McCabe's cyclomatic complexity was the first attempt to capture the attribute of testability. We have suggested that trying genuinely to measure such attributes is hopeless. Nevertheless, there is this in favour of looking for 'better' testability measures: they may at least relate directly to the task facing the software tester. If we can say that some modules exhibit a very high number of potential execution paths then this may tell an engineer engaged in white-box testing something about the magnitude of their task. In fact, product measurers do not appear to have tried very hard to forge such direct relationships with engineering tasks; we can think of only one other candidate: the attempt to measure maintainability in terms of ripple effect (see [Joiner 93] for a recent discussion).

In fact cyclomatic complexity does badly in this respect since to be precise it measures 'the cardinality of a maximum set of linearly independent walks' in a program graph; this number bears very little connection to the actual task of white box testing. It took more than ten years[1] and several pointless discussions of its ability to capture the attribute of 'complexity' before Nejmeh raised this more crucial problem with McCabe's measure [Nejmeh 88]. Prather [Prather 87] independently proposed approximately the same solution: a measure of the actual number of paths through the program graph (with a limit imposed on loop·iterations). Bache and Müllerburg [Bache 90] extended this work to a systematic catalogue of possible measures and it is this catalogue which is implemented in the QUALMS software metrics tool.

The PROMISE project made a further advance on the theory of testability measures by solving some technical problems [Bainbridge 94] concerning the loop iteration limits in [Prather 87] and [Bache 90] (some different problems have been identified and tackled by Bertolino and Marré [Bertolino 96]). These solutions may be regarded merely as theoretical preliminaries but they raised some interesting questions about the way in which product measures are defined and these have been explored as well: [Bainbridge 96, Whitty

---

[1] It would appear that, during the intervening period, only Lipaev *et al* [Lipaev 83] revisited cyclomatic complexity as a testability measure.

96a]. Meanwhile, some evidence has emerged that cyclomatic complexity is out-performed by the more advanced testability measures [Bertolino 96a] in terms of predicting testing effort (a use of the regression approach described in section 2.2).

It should be reiterated that the attribute of testability is much too complex for us to pretend that we are actually measuring it: the term 'testability measures' is merely a convenience. For example, the research mentioned above only addresses the notion of testability as being 'the difficulty of applying certain white box testing strategies'. It would nonsensical to claim that our numbers relate to testability as 'the difficulty of forcing faults to reveal themselves through testing'. This view of testability has been championed by Voas and others (c.f. [Friedman 95]) and has to do with how faults are triggered and propagated within program code. It is arguably more relevant in the present context of safety-critical applications and has therefore also been examined by the PROMISE project, with some success although again the work has been theoretical to date [How Tai Wah 96].

### PL/M-86 Front End

The first step in deploying product measurement is to have the capability of parsing source code so as to apply the measurement algorithms. In fact, this is not strictly true for early measurements such as McCabe's and Halstead's which can usually be calculated using only lexical analysis, but more sophisticated measures, such as the testability measures just described, certainly require parsing. In principle this is a straightforward task analogous to compilation but in practice the two tasks are very different; whereas a compiler can reject any program text which it cannot handle (with a suitable error message) a program analyser must work on the assumption that any code presented to it already compiles successfully. Providing a program analyser with, say, a C language front-end is therefore a matter of guessing exactly which C programs may be considered syntactically legal.

One option open to the provider of language front-ends is to adopt the compiler-writer's position and restrict use of the analyser to a specific version of the chosen language. For instance, the analyser may be guaranteed to work only for ANSI standard C or COBOL. In practice, this option is of limited value because analysers are most necessary precisely for those large systems developed using a language dialect which is non-standard and may even be specific to one organisation. Safety-critical systems, developed for specific hardware and maintained over a long period of time, are generally of this nature.

Of course, there are many high-quality program analysers in existence which must, in some way or another, circumvent the problem of parsing 'unseen' code. Our belief is that the front-ends of these tools are produced by trial and-error: a basic parser is progressively enhanced to take care of features that crop up in industrial programs. Studies such as [Bieman 88] and

[Whitty 87] have taken a more theoretical approach to the problem, assuming that a robust parser exists and concentrating on how to define an internal representation of a program as an output of the parsing process. However, such studies can only be starting point for an actual implementation.

Producing a front-end for the QUALMS software metrics tool for PL/M-86 did indeed prove to be a very expensive exercise. A basic grammar based on the manuals was used to produce a parser generated by YACC [Levine 92]. The parser was then instrumented to generate flowgraphs from PL/M programs which could be used by QUALMS for its algorithms (for example, to compute paths through the program flowgraphs). This took of the order of six personmonths of effort. But to produce a *robust* parser which worked reliably, on even a small subsystem of the software for PMS-R's SL machines, took at least another 18 personmonths. The reasons for this were: (a) as suggested above, the compiler used by PMS-R accepted a far more subtle version of the language than that defined in the manuals; and (b) it was often not possible to present QUALMS with PL/M programs defined cleanly in terms of single files of code. The necessities of maintaining a large system obviously militate against organising your software for the convenience of a software metrics tool!

Nevertheless, a robust front-end for PL/M was eventually produced [Howson 95] and used to analyse large amounts of code from PMS-R.


### *Object-oriented Software Metrics*


A further development at PMS-R since the original MSc student placements was their trialling of C++ as a possible future implementation language. It was not felt that object-oriented (OO) metrics should become the focus of the project (there was limited process data available at PMS-R for this paradigm); still the PROMISE team felt that it would be a good thing to find out what research was available in this area against the time that PMS-R might need it. Eventually, this turned out to be a more large-scale activity than expected; metrics and OO both being 'growth' areas has lead to an explosion of research in the intersection! The output of this project activity was an annotated bibliography of OO metrics publications which has been published for wider availability [Whitty 96] but which is maintained and updated on the world-wide web.

An analysis of some key issues, based on this bibliographic study, is offered in [Whitty 96d]: not much has been published on how to evaluate a particular OO method, or on how to test the cost-effectiveness of the paradigm as a whole, although for the safety-critical sector, Voas has suggested that it may militate against testability [Voas 96]. There is, however, a very large body of literature proposing new product measures, or modifications of existing ones. This imbalance in the literature is further explored in [Whitty 96c] where a comparison is made of the topics addressed by the OO metrics literature compared with the likely audiences for these topics. The comparison is based on an analysis of the possible uses of measurements at each level of the Capability Maturity Model of the Software Engineering Institute at Carnegie Mellon University [CMU 95]. This model ranks software producing organisations on an ordinal scale from 1 to 5 as shown in the left-hand column of Table 3.

The analysis in [Whitty 96c] judges that only industrial experience reports (Ind) and survey and tutorial material (Surv) are likely to be immediately useful to organisations at the initial level, where development takes place in an ad hoc manner. At the Repeatable level, where a development process of some sort is in place, the organisation may use metrics research concerning management of OO  (Man) and cost estimation and sizing (Size), as well as material concerning specific languages such as Smalltalk (Small) and C++. Only when the development process is properly defined and documented, at level three, is research concerning quality (Qual), reuse  and testing (test) likely to be of use. At the managed and optimizing levels, metrics are used as feedback into the process so that individual projects or the process itself and be optimised. Finally, at this level,  product measurements (Prod) may become useful; only at this point can there be systematic collection of process data and the ability to derive meaningful empirical results (Emp) concerning its relationship with product measures.

| Maturity Level | Relevant publication classifications | Publications having classifications relevant to level = $L_i$ | Publications having no classifications relevant to level < $L_i$ | Publications having classifications relevant to level ≥ $L_i$ |
| --- | --- | --- | --- | --- |
| 1: Initial | Ind, Surv | 87 | 87 | 267 |
| 2: Repeatable | C++, Small, Man, Size | 132 | 85 | 180 |
| 3: Defined | Qual, Reuse, Test | 93 | 37 | 95 |
| 4: Managed | Emp, Prod | 163 | 58 | 58 |
| 5: Optimizing | — | — | — | — |

Table 3: amount of OO metrics research available at each maturity level.

Unfortunately, the vast majority of organisations are at the initial level. This means, for instance, that 2/3 of OO metrics publications (namely the 180 listed in the last column of Table 3 as being relevant to level 2 or above) are not of immediate interest to the large majority of software developers. Of course, this does not say anything about any particular organisation or about any one industrial domain (procurement departments, as we mentioned earlier, may be able to make use of product measurement research without have the attributes of a level 4 organisation). On the whole, though, this research did not encourage the PROMISE project to add to the body of research in product metrics.

This research into OO software metrics has been negative in a sense and has provided little of particular relevance to the safety-critical community. Nevertheless, the annotated bibliography has proved to be a useful resource, with the web site having an average of 175 visitors per week during the final year of the PROMISE project.

### Assessment of Target Environment

An obvious pitfall for those wishing to deploy software measurement is to concentrate on the technology (which is all we have talked about so far!) to the exclusion of the target environment. The PROMISE project informally followed the **ami** method of implementing a software measurement programme [Pulford 95], in which an assessment of the environment is the first of four stages. It had originally been planned to have further MSc students working 'on the ground' at PMS-R in order to make this assessment but this proved not to be possible. Instead, a member of the project team spent several week-long periods on site at PMS-R discussing issues to do with quality assurance, development priorities and tool support [Rogers 95]. These visits proved to be very stimulating and informative for the team as a whole. Additionally, a large amount of fault data was made available and analysed for the project.

It became clear, as a result of these visits, what perhaps is not often realised by software engineering researchers: the extent to which a software development environment is a highly integrated culture with a powerful focus on a set of sector-specific priorities. For instance, the development of radiotherapy machinery is regulated by the Health and Safety Executive on behalf of the UK and by the Food and Drugs Administration (FDA) on behalf of the USA. These bodies set the agenda for quality assurance activities at an international level and it is overridingly important that organisations such as PMS-R comply with their requirements. The FDA, for example, have the power to close down all US operations of the companies they regulate and this has happened in the radiotherapy sector (not to PMS-R). By and large, therefore, the process innovations which are adopted in this sector will come from the top down, following their adoption or approval by the regulatory bodies. There is limited scope for researchers to tinker with quality assurance procedures at a local level.

It is ironic that, more than anything else, it was the dedication to a regulated process which led us to expect that here, in the safety-critical domain, was a natural home for software product metrics. In the end this very regulation militated against their uptake.

*Application of Bayesian Networks to Product Metrics*

As the PROMISE team absorbed the lessons it learned from the on-site visits it re-oriented towards a less intrusive role for product metrics. It became clear, for example, that PMS-R were not likely to relish explaining to the FDA why they had abandoned their established methods of testing in favour of the Fuchs method! The DATUM project had been examining the use of Bayesian networks for decision-making in the safety-critical domain and this work was adopted by the PROMISE project as a way of trying to integrate product data with the existing priorities at PMS-R [Neil 96].

A readable introduction to Bayesian networks is given by Heckerman and Wellman [Heckerman 95] in a special issue of *ACM Communications* devoted to the subject. Briefly, a Bayesian network can model the uncertain impact of many interacting factors on an outcome or event of interest. For example, consider the IFIP model of dependability shown in Figure 5 ([Whitty 96b] gives more explanation).

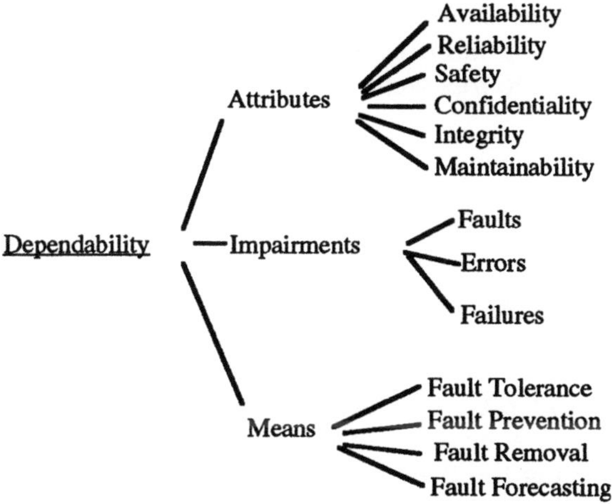


Figure 5: IFIP model of dependability.

The achievement of dependability according to this model depends, to an uncertain extent, on many 'attributes' which are present, to a greater or lesser degree, according to how successfully a number of 'means' overcome a number of 'impairments'. A Bayesian network, matching the tree above, will allow all the unknowns to be quantified subjectively (using expert judgement, for instance) and for the impact of objective inputs (such as a product measurement of maintainability) to be subjectively weighted. The result will allow, say, a quality manager to explore the outcomes of different strategies: does high maintainability militate against safety in the presence of low failure rates etc. The value of such a network for the software product measurement user lies in the way that product data can be contextualised, to a far greater extent than in any of the uses of product measurement described earlier.

There has not been time within the PROMISE project to explore properly the possibility of integrating a Bayesian network application of product measurement into the safety-critical culture at PMS-R. It is possible that this AI slant would prove acceptable whereas the more deterministic approaches described earlier would not. Nevertheless, it seems likely that there is a steeper learning curve attached to the latter application and this might militate against its success.

### Summary

The PROMISE project has made innovations in the areas of testability measurement, object-oriented software metrics, the application of Bayesian networks and language parsing. All these innovations were motivated either by the needs of the technology we hoped to put into place at PMS-R or by our perception of the needs of PMS-R. Not all have a direct bearing on safety-critical software development, although work in all these areas brought discoveries of relevance to this sector.

## 3.3 Outcome

The description of the achievements of PROMISE given above may have indicated something of the struggle which the team experienced. Problems were encountered with some of the basic theory (testability measurement). These problems were overcome after considerable effort. Applying the theory to PL/M code meant a long battle to produce a robust parser. Meanwhile, close contact with PMS-R was making it apparent that the techniques for applying the theory (listed in section 2) were not likely to be acceptable in a sector as regulated as the safety-critical domain—we had a robust theory and a robust implementation but the deployment mechanism was not robust! The theory of Bayesian networks might prove robust in time but threatened a steep learning curve. Nor was the outcome of our study of object-oriented software metrics an optimistic one: it appeared that only in special cases (which *might* include the safety-critical domain) could product measurement research expect to be useful.

It would be prevarication to pretend that after our long struggle we at last emerged victorious. The PROMISE project has not effected the successful take-up of software product measurement within the safety-critical domain. Although much code has been analysed as well as a very large body of fault data (sadly, these analyses currently remain confidential) there is, as far as we can see, no prospect of such analyses continuing beyond the funding of the project.

It remains for us to make good use of the achievements which can be marked up for the project and to try to derive general lessons from our experiences on PROMISE.

# 4 Lessons Learned from PROMISE

I think that software product measurement technology is somewhat akin to software unit testing technology: some appealing theory has been embodied in tools which appear to be quite slick. But the tools often cannot cope with real industrial code without an inordinate investment in customisation; the testing techniques often bear little relation to what is actual industrial practice; and there is often a culture which militates against re-orientation towards the new techniques.

Although measuring software products seems, on the face of it, to be a sensible thing to do, we would suggest on the basis of PROMISE that only if the following are satisfied is it likely to be worthwhile:

(1) the organisation concerned should be at level 4 or 5 on the SEI's capability maturity model; if there is reason to doubt that this is the case, then there must be a special reason and special support for product measurement (as may be the case in the procurement domain, for example);

(2) robust parsers should already exist for the languages concerned, or it must be anticipated that the use of product measurement will ensure a return on the significant investment required to produce such parsers;

(3) there must be a culture conducive to trialling and adopting radically new methods of quality assurance.

It would appear that none of these conditions were fully met in the case of the PROMISE project. We believe it was predominantly because of this that the project failed in its principal aim of deploying product measurement in the safety-critical domain. However, we hope that there is room for discussion about this and that the PROMISE project can be of some use to the safety-critical domain in contributing to this discussion. Certainly, we can think of other possible reasons for our failure. We may have been over-ambitious in hoping to deploy state-of-the-art product measurements; it might have been better to concentrate on getting started with McCabe's cyclomatic complexity (which would only have required lexical analysis). We spent a lot of time analysing PMS-R's process data; it might have been better to start with product profiling which requires none. Our contacts at PMS-R were mainly experts from the development side; it might have been better if we had more contact early on with experts in matters relating to FDA and HSE regulation.

We hope we have been frank enough about the project to allow readers to form their own opinions on these issue. We hope also that we have been fair to ourselves and been frank without being too negative. We learned a good deal about the technology of product measurement itself. The scientific community benefits from such progress, often in ways quite unforeseen by the originators. Systems written in PL/M and similar 'legacy languages' will be around for a long time, perhaps particularly in the safety-critical domain with its historical background in hardware and its cautious approach to change. Everything we learn about such systems can have an impact, directly or indirectly, on our ability to deal with their idiosyncrasies.

## Acknowledgements

PROMISE was supported by the EPSRC as part of the EPSRC/DTI programme of research and development in safety-critical systems (grant no. J18880). This paper was written while on sabbatical with LAAS du CNRS, France. I am grateful to them for their hospitality and to the Royal Academy of Engineering who supported my visit with an Engineering Foresight award. Much of the work described here was carried out by my colleagues on the PROMISE project: John Bainbridge, Eddie Howson, K.S. (Roland) How Tai Wah, Lasitha Leelasena, Martin Neil and Jennie Rogers at South Bank University and Kevin Bray and Martyn Must at PMS-R.

## References

[Bache 90]        Bache R. And Müllerburg M. Measures of testability as a basis for quality assurance. Software Eng. J. 1990; 5 (3): 86–92.

[Bainbridge 94]   Bainbridge J.R. Defining testability metrics axiomatically. Software Testing, Verification and Reliability 1994; 4 (2): 63–80.

[Bainbridge 96]   Bainbridge J.R. and Whitty R.W. Recent research in program flowgraph theory. in Mitchell C. (ed.) Applications of Combinatorial Mathematics. Oxford University Press (to appear).

[Bertolino 96]    Bertolino A. and Marré M., How many paths are needed for branch testing? to appear in J. Systems and Software 1996.

[Bertolino 96a]   Bertolino A., Mirandola R. and Peciola E. A case study in branch testing automation. In Bologna S. and Bucci G. eds. Achieving quality in software: Proceedings of the 3rd International Conference on Achieving Quality in Software (AQuIS'96). Chapman and Hall, London, 1996: 369–380.

[Bieman 88]       Bieman J.M., Baker A.L., Clites, P.N., Gustafson, D.A. and Melton, A.C. A standard representation of imperative language programs for data collection and software measures specification. J. Systems and Software 1988; 8: 13–37.

[Burr 96]  Burr A. and Owen M. Statistical methods for software quality: Using metrics for process improvement. International Thomson Computer Press, London 1996.

[CMU 95]  Carnegie Mellon University, Software Engineering Institute. The Capability Maturity Model: Guidelines for Improving the Software Process. Addison-Wesley, Reading, MA, 1995

[Courtney 93]  Courtney R.E. and Gustafson D.A. Shotgun correlations in software measures. Software Eng. J. 1993; 8 (1): 5–13.

[Fenton 92]  Fenton N.E. When a software measure is not a measure, Software Eng. J. 1992; 7 (5): 357–362.

[Fenton 96]  Fenton N.E. and Pfleeger S.L.  Software metrics: a rigorous and practical approach, 2nd edition, International Thomson Computer Press, Andover, Hampshire, UK, 1996.

[Friedman 95]  Friedman M.A. and Voas J.M. Software assessment: reliability, safety, testability. Wiley, New York, 1995.

[Halstead 77]  Halstead M. Elements of software science. Elsevier, New York, 1977.

[Hamer 82]  Hamer P. and Frewin G. Halstead's software science: a critical examination. in Proc. 6th Int. Conf. on Software Eng., Tokyo, Japan, 1982; 197–206.

[Heckerman 95]  Heckerman D. And Wellman M.P. Bayesian networks, Comm. ACM 1995; 38 (3): 27–30.

[Howson 95]  Howson E.F. PL/M-86 front-end for QUALMS. MSc Dissertation, University of Greenwich, 1995.

[How Tai Wah 96]  How Tai Wah K.S. Fault coupling in finite functions. Submitted to ACM Trans. on Software Eng. and Methodology.

[Joiner 93]  Joiner J.K. and Tsai, W.T. Ripple effect analysis, program slicing and dependence analysis. Technical Report TR 93-84, University of Minnesota, Minneapolis MN 55455, 1993.

[Leveson 93]  Leveson N.G. and Turner C.S. An investigation of the Therac-25 accidents. Computer July 1993: 18–41.

[Levine 92]  Levine J.R., Mason T. and Brown D. LEX and yacc. O'Reilly & Associates, 1992.

[Lipaev 83]  Lipaev V.V., Pozin B.A. and Stroganova I.N. Complexity of program module testing. Programming and Computer Software 1983; 9: 332–337.

[Lockhart 93]  Lockhart R. On statistics in software engineering management. Software Quality Journal 1993; 2: 49–60.

[McCabe 76]  McCabe T. A complexity metric. IEEE Trans. Software Eng. 1976; SE-2 (4): 308–320.

[MacKenzie 95]   MacKenzie D. Computer-related accidental death: an empirical exploration. Science and Public Policy - Computer Safety 1994; 21 (4): 233–248.

[Nakassis 92]   Nakassis C. Software metrics data collection and analysis. MSc Dissertation, South Bank University, 1992.

[Neil 96]   Neil M., Littlewood B. and Fenton N. Applying Bayesian belief networks to system dependability assessment. in Redmill F. and Anderson T. (eds.) Safety-critical systems: the convergence of high tech and human factors. Springer-Verlag, London, 1996: 71–94.

[Nejmeh 88]   Nejmeh B.A. NPATH: a measure of execution path complexity and its applications. Comm. ACM 1988; 31 (2): 188–199.

[Prather 87]   Prather R.E. On hierarchical software metrics. Software Eng. J. 1987; 2 (3): 42–45.

[Pulford 95]   Pulford K., Combelles A. and Shirlaw S. A quantitative approach to software management: the ami handbook. Addison-Wesley Longman, Wokingham, UK, 1995.

[Rogers 95]   Rogers J. Safety needs communication. Safety Systems. The Safety-Critical Systems Club Newsletter, May 1995, 4 (3): 14.

[Shepperd 88]   Shepperd M. A critique of cyclomatic complexity as a software metric. Software Eng. J. 1988; 3 (2): 1–8.

[Voas 96]   Voas J.M., Object-oriented software testability. in Bologna S. and Bucci G. (eds.) Achieving quality in software: Proceedings of the 3rd International Conference on Achieving Quality in Software (AQuIS'96). Chapman and Hall, London, 1996: 279-290.

[Whitty 87]   Whitty R. Modelling sequential processes for complexity measurement.  Paper 7 in Structure-based Software Metrication,. CSSE, South Bank University, 1987: 185–196.

[Whitty 96]   Whitty R.W. Object-oriented metrics: an annotated bibliography. SIGPLAN Notices 1996; 31 (4): 45–75.

[Whitty 96a]   Whitty R.W. Computing. In Wilson R. And Beineke L. (Eds.) Graph Connections. Oxford University Press (to appear).

[Whitty 96b]   Whitty R. Quality attributes in critical systems: a report on a talk given at the IEE by Mario Barbacci. Safety Systems. The Safety-critical Systems Club Newsletter 1996; 5 (2): 7–9.

[Whitty 96c]   Whitty R.W. Applying process maturity assessment data to evaluate research in object-orient software metrics. Paper presented at the 1996 Bournemouth Metrics Workshop, University of Bournemouth, April 1996.

[Whitty 96d]   Whitty, R.W., Object-oriented metrics: a status report. Object Expert January/February 1996; 39-44.

# Multi-disciplinary Projects and Technology Exchange - the SEMSPLC[1] Experience


Audrey Canning
ERA Technology Ltd.
Leatherhead, Surrey


**Abstract**


In 1991 the UK Government awarded the project 'Software Engineering Methods for Safe Programmable Logic Controllers (SEMSPLC) to a consortium of fourteen organisations. The organisations were drawn from several application sectors including process control, machinery control, software engineering and food manufacture and from across the supply chain - including end-users, systems integrators, tools suppliers and regulatory authorities. The project aimed to bring together the experiences of the many different participants for the mutual benefit of all. Whilst the widely differing experience and perspectives of the partners brought many challenges, the end result has been a document which has the full support of both practitioners and theoreticians. Moreover, both the research community and the industrialists have gained new insights into the practice of software engineering. This paper summarises the insights gained and outlines the ways in which practical technology transfer between widely differing communities was achieved.


# 1      Introduction

In September 1996 the Institution of Electrical Engineers (IEE) published the document *"SEMSPLC Guidelines - Safety-Related Application Software for Programmable Logic Controllers"* [IEE96]. The document represented the results of a five year study. The first four and half years, supported by the UK Government[2], involved the development of software engineering standards (informed by theory and industrial practice) followed by their representative use on two trial projects. The final six months was sponsored mainly by the participating companies and the IEE itself to gain industrial feed back through consultation with trade associations and leading practitioners within the field.

---


[1]      The SEMSPLC project partners were ERA Technology, British Gas, ICL, LDRA, Servelec, University of York and YSE. The project's sponsors were CEGELEC Projects, Cincinnati Milacron UK, Eutech Engineering Solutions, Gerrard Software, Health and Safety Executive, ICI Engineering Technology, IDEC, Nestle UK and Nuclear Electric.

[2]      Funded as project IED4/1/9019 under the joint DTI/EPSRC Safety Critical Systems Advanced Technology Programme (SCSATP).

The technical results of the project have been very successful. Both demonstrator trials have independently confirmed that use of the Guidelines leads to a reduction in the number of errors leading to unsafe Programmable Logic Controller (PLC) behaviour. Further, all three organisations which have used the Guidelines in practical trials were keen to use the practices in their day-to-day activities (subject to competitive constraints). With regard to technology development a number of observations have been made which in some cases challenge the assumed wisdom held in traditional software engineering practices. All organisations involved in the project agree that their awareness of software systems engineering  and safety engineering has been enhanced.

By the end of the project the team had established a very high level of cohesion, one example being the continued involvement of team members in exploitation through a series of public seminars. Nevertheless, many different viewpoints were adopted by partners leading to significant debates and, from time to time, apparently irreconcilable technical positions. In the following sections the issues surrounding the technical management of the project are discussed which, it is hoped, will provide insight into how technology transfer may be achieved among partners with very different interests and backgrounds.

## 2      The Need for the Project

The concept for the SEMSPLC project has its origins in industrial work carried out in 1985. At this time ERA Technology became involved in independent safety assessment and auditing for the off-shore oil and gas industry against the, then draft, HSE guidelines for the use of programmable electronic systems (PES) in safety related applications [HSE87a, HSE87b]. The industry was starting to deploy unmanned platforms within the North Sea and, for commercial reasons, it was appropriate to use programmable electronic systems for remote fire and gas detection and emergency shutdown. Whilst ERA's work was limited to the safety assessment of the systems,  it was very clear that significant commercial difficulties were experienced, both in achieving adequate functionality and in adhering to rigorous schedules during factory acceptance test and during on-site commissioning. A further observation was that the architecture of the particular form of computer system used, the Programmable Logic Controller, greatly influenced the software development methods used and called for specific features within the safety checklists. One notable point was that the specification and development of PLC software tends to be dominated by its input/output relationships with the external world and functionality is derived from these relationships. Thus, when structured techniques were used these appeared to be largely a waste of time, being limited to the reproduction of the Cause and Effect Logic prepared by the customer, but in a less appropriate notation. Programming errors were frequent, but were regarded largely as "simple" typing errors, since the most prevalent error was misuse of data tags within the large global database which embodied the program complexity.

Over the succeeding five years several industrial safety shutdown systems were assessed, and the observation was made that the quality and commercial success of the projects appeared to be very dependant on the skills of individual programmers. Moreover, there was very little management experience in predicting which project would be successful. Clearly this observation is significant, not just for commercial reasons, but also when one considers the integrity requirements of the safety systems being used.

When the UK government announced the Safety Critical Systems Programme in 1990 the industrial need for commercially developed programmable electronic safety systems appeared strong. A letter was circulated to industry describing the nature of the perceived problems and inviting both users and developers to become involved in a pre-competitive collaborative research project to address this issue. The high level of interest confirmed the viewpoint that this was an important area for suppliers and users alike and hence the SEMSPLC partnership was formed.

# 3 The Form of the Project

Clearly, for a collaborative project to be successful, the commercial aims of the participants must not be in conflict. For this reason it was determined to limit the shared deliverables to items where the chief gain to participants could be achieved through wide publicity. In addition, since most participants were investing substantial sums of money and effort in the project, additional deliverables were needed depending on the nature of the individual company's businesses. In the event a "methods" document, intended for wide publication, was selected to be the main shared deliverable, together with a public report evaluating the method. Individual deliverables took the form of prototype tools which could be used to demonstrate the techniques developed on the project. However, the primary exploitation routes for individual deliverables would be through the responsible partner developing the ideas within the consortium. Partners involved in using the technology would benefit primarily through the opportunity to try out the ideas in practice and through developing internal methods in line with the emerging public method.

Given the business profile of the partners, and their commercial needs, the principal objectives of the SEMSPLC project were:

1.  to establish a software engineering method suitable for the development of safety-critical PLC based programs in an application language suited to industrial control; this aim fully supports the need for a pre-competitive focus for the project targeted on achieving real benefits for industrial safety system development;

2.  to define conformance standards for tools which could support the application of the method; this aim allowed partners the opportunity to

demonstrate technology capabilities whilst still retaining an exploitation edge within their individual business interest areas.

A further need within the project was to ensure a strong technology focus. In particular, the interests of the individual project team members (as opposed to their company's commercial interests) were, by-enlarge, technically motivated. The actual technical drivers varied widely between the different groups - application of both business process modelling and of formal methods were key topics in the early 1990s and several team members were motivated to understand these areas better. A strong driver for the project manager was to understand the mapping between control theory and the discrete models used within digital systems. Other research issues, such as the ability to "measure" the level of safety, needed to be addressed in order to demonstrate progress against the principal project objectives. A further research topic, that of the re-use of safety critical software, arose from the pragmatic observation that industrial practice as used in industrial safety systems such as these based on PLCs could not be handled fully through the emerging standards of the day.

The form of the work plan, within the constraints of achieving the business objectives, mainly reflected the technical interests of the individual team members. For this reason the project was divided into three phases as outlined below.

- The *first phase* (nine months) focused on the individual technical study of various topic areas so as to bring together as much information and experience as possible.
- The *second phase* (nine months) was devoted to establishing the back-bone of the project deliverables based predominantly on the use of business process modelling and formal methods.
- The *third phase* (two and a half years) was given to practical experimentation and feedback from industrial participants. This phase was deliberately long term allowing sufficient flexibility to fit in with the industrial constraints of the partners. In particular it was intended to link the project work with "real" work, since partners found the actual timescales for industrial work somewhat difficult to predict. During this third phase the technical partners were to concentrate on refinement of technical ideas based on insights gained from the industrial experiments.

## 4 Working in a Collaborative Environment

The project started well, with the production and agreement of the collaboration procedures, project reporting, liaison and configuration management procedures and the project's information dissemination procedures. A fundamental principal underpinning the project was that each member could pursue individual commercial interests, but that such commercial interests must be declared and agreed at the project steering group level before affecting other members of the group. For example, any member could introduce background IPR, but that this must be declared in advance; any member could propose to publish documents concerning their own work (and indeed this was encouraged within the project as one of the important dissemination aims of the DTI) but that before doing so the agreement of the steering group must be obtained. Monitoring of such matters proceeded as a matter of routine at steering meetings and through regular items in progress reports.

The technical work also commenced well, with each partner taking responsibility for individual tasks within the Phase 1 work plan. Briefing papers and reports were produced in ample numbers and the project came together to review and approve the papers. The review process did not go well. In particular, there proved to be very little common ground between collaborators so that solutions proposed from one viewpoint would appear totally unrealistic from another viewpoint. The problems were compounded by terminological differences which made it difficult to discuss individual problems and to convey the nature of the concerns. Typical examples of the of the problems experienced are given below.

1.  Differing perspectives between those engaged in development of safety systems (where the focus is primarily on ensuring unsafe situations cannot occur) and those primarily concerned with producing adequate systems on time and to budget. In the latter case the drive is more towards repeatable quality within controlled costs and timescales. With hindsight it is suggested that many conventional software engineering practices have developed more as a result of commercial considerations than safety considerations and that this tends to confuse the novice practitioner, who will accept the techniques as necessarily producing "good" software.

2.  Differing perspectives between those engaged in industrial engineering practice compared with those involved in developing ideal technologies. One particular technical sticking point was related to an ideal requirement to radically redesign the foundation of the PLC kernels in order to assure timing properties when compared with the reality of commercial success in a multi-billion pound commercial market.

3.  Differing technical experiences in terms of the software development methods used. In particular the difference between PLC programming and conventional software development practices is so great that through discussion it was not possible for the two communities to gain a common understanding of the advantages and disadvantages of one another's technologies. The problem was compounded by each member claiming a certain "moral highground" in areas where their beliefs were particularly strong, thus blocking the effective transfer of information.

In order to resolve these issues four key strategies were adopted.

1. The process modelling phase of the project was re-directed, not to construct an ideal process but rather to capture, analysis, understand and reconcile the actual practices ongoing in each of the seven industrial process control organisations; the results of the activity were surprising for two reasons:

   (a) the ease with which the different organisation's activities could be reconciled, indicating that there was a good consensus on the methods currently in use in the process control industry, and
   (b) the way in which the exercise made explicit those parts of the process which were not at that time captured in development standards, but which were carried out by a significant number of the organisations.

2. Cross-partner task groups were set up to establish how the methods and techniques identified during the first phase of the project could be used to improve the existing processes with respect to the safety of the final product. In most cases the task group leader was selected from the organisation with a task responsibility in the particular area however, in a few cases the task group leader was selected on the basis of their management skills and their ability to reconcile widely dissimilar views. Each collaborator was required to be part of a number of task groups and each task group was required to incorporate a number of different types of organisation. In practice six disciplines were established; formal methods, timing, languages, hazard analysis, metrics and testing. An important activity within each task group was the arrangement of practical demonstrations and "teach-ins" in order to gain an understanding of the perspectives of the different organisations

3. The evolution of the SEMSPLC method was divided into two activities:

   (a) evolution of the existing practices as described through the process models in the direction of the ideal recommendations, without requiring a major change in the investment practices of the PLC market in terms of the existing hardware and support environments, and
   (b) definition of additional ideal requirements incorporating more radical advancements of technology should it be proposed to use PLCs in very critical safety areas.

4. A formal deliverable review procedure was introduced, essentially based on peer review to ensure that the team became self policing.

The results of the changes in strategy, from both technical and project management viewpoints, were very rewarding. During one notable practical session the conventional software engineers attempted to use PLC programming techniques and found that experienced PLC programmers utilise very sophisticated, albeit alien, skills to achieve elegant programming solutions. Similarly, during certain practical testing activities the PLC practitioners were shown errors within their program which they had believed could not exist (they later enthusiastically went on to apply the same techniques to find further errors in their own programs!). The result was that both communities gained far deeper insight into the technology of the other and became far more willing to consider the uptake of the complementary technology. For the software research community new application areas within the industrial community were perceived. Within the industrial community there was a far greater willingness to try out the theoretical techniques.

Another curious effect observed on the project were the dynamics of the team relationships. In particular, during full project review meetings individual members could be seen first siding with one grouping and then with another as different issues were discussed. Each sub-group appeared to reach an internal understanding on its technology which would be fiercely defended in the wider group. However the cross grouping  relationships were such that for another topic a very different alignment of team members was seen. In general this had the effect of fostering a very supportive relationship within the team. The team cohesion continued to be clearly observable throughout the IEE industry review.

The practical work on the project also led to two challenges to conventional software engineering. The first, concerning the conventional wisdom on test practices, arose from the observation that conventional test coverage metrics yield little information when applied to PLC programs. In particular, a PLC program may have a simple control flow structure, the complexity being embodied in the data, and a simple test case can achieve full coverage of the control flow path. Clearly, the complexity is still present and an enhanced test method for this type of program was found to be a form of mutation testing.

The second challenge relates to the beliefs surrounding software engineering methods. In particular the conventional wisdom is to use forms of data analysis (data flow, object orientation etc.) to define the high level design of the software. This approach is recommended within the SEMSPLC guidelines and was implemented within the trial projects. The results were surprising. The size of programs incorporating software engineering methods were approximately twice the size of those of the traditional programs and, when test coverage metrics were reviewed, a lower level of conventional white box coverage metrics were revealed compared with that achieved for the original industrial methods. At the same time the coverage of requirements, the number of errors found and the rework costs of the programs were much less than those using the industrial methods. Several points arise from these observations.

1.   Should different types of test metric be devised for different types of software development method? In particular, it was concluded that coverage

of the SEMSPLC method would be better addressed through measurement of the coverage of the states within the program.

2. What is the pertinent trade-off between development method, program size and complexity? In particular, the size of programs in the SEMSPLC project did not appear to relate directly to complexity of the system states.

3. How should different types of requirement within a program be represented? In particular, for the industrial program techniques, the interface driven requirements produce a simple view of the program's functionality. However, superimposed upon those requirements are a set of additional requirements (such as overrides, self test etc.) which do require an additional form of representation such as those provided by software engineering methods.

4. How do these findings relate to other types of industrial program which also are often predominantly data driven?

## 5      Practical Project Results

The practical results of the demonstrator trials do confirm that the SEMSPLC method can assist in producing software which is less likely to suffer from unsafe defects [ERA95]. All three organisations involved in using the Guidelines have been keen to adopt their use in day-to-day activities (two of the three are developing internal practices based on the method while the third notes that it will be necessary to achieve industry pressure to make the practices commercially viable). The industry feedback achieved through the IEE was also favourable; no major comments were raised that were within the scope of the original work programme, although a significant number of minor corrections were made to the document as a result of the industry comment. There is significant interest in exploitation from the partners, with four of the seven partners having specific marketing objectives directly associated with the Guidelines (and generally complementary to one another), two having the potential to exploit the technology once the market can support investment costs, and only one feeling that the predominant benefit has been limited to enhanced technology understanding. In addition, it is hoped that further benefit will accrue through the links with the IEE and with other standards organisations.

## 6      Conclusions

This paper has discussed some of the practical experiences gained during the conduct of a large, successful, multi-disciplinary, multi-sectorial collaborative research project. Key components of the success are attributed to:

1.  the simultaneous focus on satisfying the industry need (the task) the technology interests of team members (the individual) and achieving practical interdisciplinary working groups (the team);

2.  the early recognition of the need to manage and reconcile the commercial interests of the different parties, reflected both in the deliverables structure and in the working procedures;

3.  the development of a framework to accommodate the different technical viewpoints through subdivision of the SEMSPLC Guidelines into two levels of recommendations;

4.  use of the team spirit to police the working relationships and quality between the individual organisations and team members, and

5.  involving the enthusiasm and commitment of outside bodies such as the HSE and the IEE.

## Acknowledgements

The author would like to thank Stephen Hatton, Roger Shaw and the SEMSPLC partners for their help in producing this paper.

## References

[ERA95]   ERA. Appraisal of the SEMSPLC Demonstrator Projects, Technical Report SEMSPLC 9019/3A/ERA/8000/R/1, ERA Technology, May 1996.

[HSE87a]   *HSE. Programmable electronic systems in safety related applications: Part 1 An introductory guide*, HMSO, 1987.

[HSE87b] HSE. *Programmable electronic systems in safety related applications: Part 2 General technical guidelines*, HMSO, 1987.

[IEE96] IEE. *SEMSPLC Guidelines Safety-related application software for programmable logic controllers*, IEE, 1996.

# AUTHOR INDEX